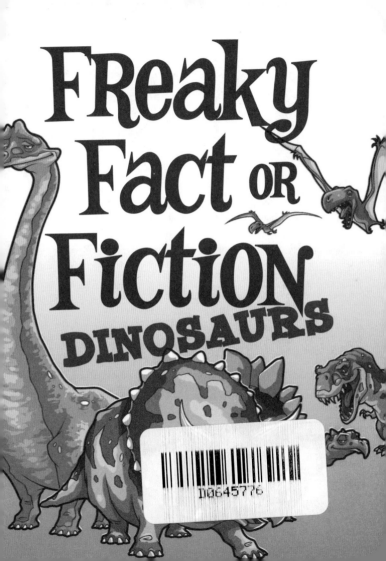

Freaky
Fact or
Fiction
DINOSAURS

hinkler

Published by Hinkler Books Pty Ltd
45–55 Fairchild Street
Heatherton Victoria 3202 Australia
www.hinkler.com.au

© Hinkler Books Pty Ltd 2011

Author: Bill Condon
Editor: Suzannah Pearce
Copyeditors: Helena Newton and Susie Ashworth
Design: Diana Vlad and Ruth Comey
Cover Illustration: Rob Kiely
Illustrations: Brijbasi Art Press Ltd
Typesetting: MPS Limited
Prepress: Splitting Image

The publisher has made every effort to ensure that the facts and figures in this
book are correct at the time of publication. The publisher is not responsible for the
content, information, images or services that may appear in any books, journals,
newspapers, websites or links referenced.

ISBN: 978 1 7418 5256 1

Printed and bound in China

Freaky Fact or Fiction

They say that truth is stranger than fiction . . . but can you tell the difference?

This book contains over 200 strange and interesting stories about dinosaurs. Most of these are true, but some are tall tales; it will take a real dinosaur expert to spot the difference.

Quiz yourself, your parents, your little sister or your best friend. You can record your answers by ticking one of the circles at the bottom of each page. Then, to check whether you were right, turn to the answers section at the end of the 'facts'.

For extra fun, we've included our sources at the very end of this book. If you want to read more about dinosaurs, or if you just want to double-check a fact that sounds crazy, the sources are a good place to begin your research.

You can start anywhere in the book and read the facts in any order. Whatever you do, get ready for hours of *Freaky Fact or Fiction* fun!

1 There were many different types of dinosaurs. Dinosaurs are usually classified by what they ate. Carnivores were meat-eaters; herbivores were plant-eaters; omnivores ate both meat and plants. During the time of the dinosaurs, there were also avian reptiles (those that could fly) and aquatic reptiles (those that lived in the water). The most dangerous of the dinosaurs were the carnivores. This group included the best-known dinosaur of all, the *Tyrannosaurus rex* (tie-RAN-uh-SAWR-us rex) – popularly known as the *T. rex*.

 **✓ FACT** OR **FICTION**

Dinosaurs

Dinosaurs first appeared on Earth some 250 million years ago, in what is called the Triassic period. They were the kings of the beasts then, and remained so through the Jurassic and Cretaceous periods. They became extinct about 65.5 million years ago. We know so much about them today because throughout the world dinosaurs have left their mark in the shape of fossils, such as bones – lots and lots of bones – and also skin, claws and teeth. New information comes to light about them every year and for the millions who are fascinated by them, they remain the kings of the beasts.

 FACT **OR** **FICTION**

Freaky Fact or Fiction

3 Scientists believe that in the time of the dinosaurs all the Earth's continents were joined in one super-continent known as Pangaea. This is a Greek word that means 'all lands'. Over millions of years, Pangaea broke apart and the various pieces 'drifted', before finally settling into the shape of the world we know today. The break-up meant that dinosaurs were transplanted to different lands. One reason why dinosaurs are found in so many weird and wonderful shapes and sizes is that they had to adapt and evolve to suit the climate and terrain of their new homes.

✓ FACT   OR   FICTION

4

In 1842, the word 'dinosaur' (Greek for 'terrible lizard') was coined by Sir Richard Owen (1804–1892), who also founded the Natural History Museum in London, England. Owen is also famous for identifying the existence of New Zealand's giant extinct bird, the moa. Owen even gave Queen Victoria's children biology lessons. However, he was a controversial figure in his day, with many claiming that while naming the superorder Dinosauria and identifying the moa, he failed to acknowledge the work of others in the field.

✓ **FACT**   **OR**   **FICTION**

Freaky Fact or Fiction

5 Do you collect things like owls, footy cards or stamps? The Heyuan Museum in China collects dinosaur egg fossils. The count is over 10,000 and the museum is still collecting. This is the biggest collection of dinosaur egg fossils in the world. By the way, all of the eggs are much bigger than bird eggs (as you might have guessed).

 ✓ **FACT** **OR** **FICTION**

6

Dinosaur remains have been found in polar regions. In 1986, the first species of Antarctic dinosaur was found on Ross Island. It was a species of *Antarctopelta* (an-TARK-tow-PEL-ta), an ankylosaurid (an-KIE-luh-SAW-rid) dinosaur. These dinosaurs had heavy armour all over their bodies – probably to protect them. However, they didn't need a fluffy coat to keep warm. When dinosaurs were in Antarctica it wasn't covered in ice!

 FACT **OR** **FICTION**

Freaky Fact or Fiction

7 Professor Kylie Minno at England's Rocksford University has started teaching a course on assembling dinosaur bones for beginners. 'A child can bang together a heap of dinosaur bones very quickly,' she said. 'They just have to remember that the toe bone's connected to the foot bone, the foot bone's connected to the ankle bone, the ankle bone's connected to the leg bone, the leg bone's connected to the knee bone, the knee bone's connected to the – well, I won't go on – but you can see how easy it is.'

✓ FACT OR FICTION

8

Elliot is the nickname of Australia's largest dinosaur. Originally weighing as much as five African elephants, Elliot is about 95 million years old. His remains were found near Winton, Queensland, in 1999. Elliot is a sauropod (SAWR-uh-pod), a four-legged plant-eater with a long neck and a disproportionately small head compared to the rest of its body. His remains were found over a space the size of seven football ovals.

✓ **FACT**　　　　**OR**　　　　**FICTION**

9

eading scientists now agree that a massive asteroid that fell to Earth 65.5 million years caused the extinction of dinosaurs. The asteroid that slammed into Mexico is said to have been one billion times more powerful than an atomic explosion. It created clouds that blotted out the sun for ten years. Scientists say that the asteroid crash resulted in 90 per cent of life on Earth being wiped out.

 ✓ FACT **OR** **FICTION**

10

T*yrannosaurus* (tie-RAN-uh-SAWR-us) Sue is the world's most complete *T. rex* skeleton. It is named after Sue Hendrickson. In 1990, Ms Hendrickson discovered the 85 per cent complete, fully-grown specimen in the US state of South Dakota. It is estimated to be approximately 67 million years old. Sue is 12.8 m (42 ft) long – the length of a bus. Sue is now a permanent exhibit at the Field Museum in Chicago.

✓ **FACT**   OR   **FICTION**

Freaky Fact or Fiction

11 In 1906, the fossil remains of two previously unknown dinosaurs were found in a cave in India. They were about the same size as a cow. Scientific studies revealed that their heads had been covered in sores. Because of this they were given the name *Soreasaurus* (SAW-ree-a-SAWR-us). It was first believed that their sores had come about because they had poor eyesight and were constantly hitting their heads. A more recent theory is that the sores may have been caused by prolonged kissing!

✓ **FACT**   **OR**   **FICTION**

● ●

12

The dinosaur commonly called *Brontosaurus* (BRON-toe-SAWR-us) is really an *Apatosaurus* (a-PAT-uh-SAWR-us). Othneil Charles Marsh named *Apatosaurus* in 1877. Two years later, he found a more complete skeleton, which he thought belonged to a new genus (animal or plant group). He named this *Brontosaurus*. Many years later, it was discovered that the *Brontosaurus* was really from the same genus as the *Apatosaurus*, but was a more mature specimen. Since *Apatosaurus* was the first name given, it is the correct term, although many people still refer to it as *Brontosaurus*.

 FACT **OR** **FICTION**

13

The term 'dinosaur' stems from two Greek words: 'deinos', meaning 'terrible, or fearfully great'; and 'sauros', meaning 'lizard'. They were the dominant land animals for more than 160 million years. Although many of them were of gigantic size, some of the earliest species, such as the *Oldensaurus rex* (OLD-en-SAWR-us rex), were no bigger than a domestic dog. In 1962, fossil remains of the *Oldensaurus rex* were found in Italy. Small round stones were heaped near the remains. This has led scientists to think that the *Oldensaurus rex* may have been a primitive ancestor of the dog breed olden retriever.

✓ **FACT**   **OR**   **FICTION**

Dinosaurs

14

A research team at Rocksford University, headed by Dr John Smythe, has found evidence that the iceberg that collided with the Titanic in 1912 may have partly consisted of dinosaur bones. Dr Smythe said small bone samples found by divers at the wreck site are 'quite possibly' those of dinosaurs. He said that unlike other parts of the world where dinosaur skeletons would in time be buried in the ground, dinosaurs from the Antarctic lay on the ice and eventually drifted to sea as icebergs.

 ✓ FACT **OR** **FICTION**

Freaky Fact or Fiction

15

*V*elociraptor (vel-os-i-RAP-tor), often just called raptor, was a turkey-sized animal commonly found in dry desert regions. Its remains were first discovered in Mongolia in 1922. It was a meat-eater and a fierce fighter. It had a long, stiffened tail with huge claws on its back feet, which it most likely used to attack or cling to its prey. *Velociraptor* had feathers and is closely related to today's modern birds.

✓ **FACT** OR **FICTION**

16

The name *Tyrannosaurus rex* (tie-RAN-uh-SAWR-us rex) means 'tyrant lizard king'. It is thought to have been able to eat up to 230 kg (500 lb) of meat in one bite. Scientists estimate it bit down on its prey with huge force. Most humans bite with a force of about 1207 kPa – kilopascals (175 psi – pound-force per square inch). However, researchers say the dinosaur probably bit with a force of up to 20,760 kPa (3011 psi).

 FACT **OR**   **FICTION**

Freaky Fact or Fiction

17 Compared to the rest of its body, *T. rex* had very short arms. They do not appear to have had any practical use, except perhaps to hold its struggling prey close to its chest during an attack. It is also possible that the arms had no use at all and, over millions of years, they would have completely disappeared. Snakes once had limbs, but gradually adapted to their present, sleeker form, which is much more suitable for their environment. This process of change over long periods of time is called evolution.

✓ FACT   OR   FICTION

18

The first recorded finding of dinosaur bones was made in 1862 by an Italian farmer, Alphonse Migarella, who made the discovery while ploughing a field near Florence. In a freak accident, the protruding hind leg of a *Stegosaurus* (STEG-uh-SAWR-us) pierced the sole of Migarella's left shoe, wounding his foot, and blood poisoning set in. Despite the best efforts of doctors, Migarella died a week later, making him the only known human victim of a *Stegosaurus*.

✓ **FACT**   **OR**   **FICTION**

19

Artists work with paleontologists to try to figure out how dinosaurs really looked. In the 1800s, a British artist, Benjamin Waterhouse Hawkins, created the first life-sized sculpture of a dinosaur called *Iguanodon* (ig-WAHN-o-don). Waterhouse worked with paleontologist and zoologist Richard Owen who worked out the probable size and shape of the dinosaur, based on existing parts of its skeleton. Hawkins' dinosaurs were a major attraction at the London Crystal Palace Exhibition of 1853 to 1854. Before the *Iguanodon* model was finished, a dinner party was held inside it!

 ✓ **FACT**   **OR**   **FICTION**

20

Australia has its own lizard of Oz. Called the *Ozraptor* (oz-RAP-tor), it is Australia's oldest known dinosaur, probably living well over 100 million years ago. Some students found part of its tibia (leg bone) in the 1960s. For a long time it was thought that the bone was from an ancient turtle. It took another 30 years before anyone realised that it was a dinosaur bone. Its name means 'Australian plunderer'. It was a meat-eater that wasn't very big. It was about 3 m (9.8 ft) long and weighed approximately 45 kg (100 lb).

✓ FACT   OR   FICTION

Freaky Fact or Fiction

21 One of the most amazing pterosaurs, or flying reptiles, was *Pteranodon* (TER-an-o-don). This was a huge animal with a wingspan of 7 m (23 ft). This is wider than any modern or extinct bird. Despite this, *Pteranodon* had a very light body – about 17 kg (37.5 lb). It must have been able to flap its wings and fly, as it had joints and muscles found in flying animals. It also had hollow bones filled with air.

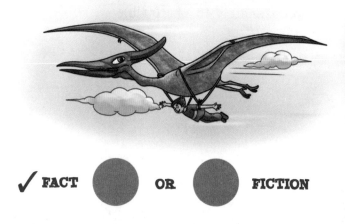

✓ FACT   OR   FICTION

22

Protoceratops (pro-tow-SAIR-a-tops) had a sharp beak and a bony frill on the back of its neck. It also had thick bones on top of its nose and over its eyes. It walked on all fours and appears to have lived in large groups. In 1922, several skeletons were found in Mongolia. Skeletons of young *Protoceratops* have also been found, some still inside their eggs.

✓ **FACT**   **OR**   **FICTION**

23

Fossils of *T. rex* prey suggest it crushed and broke bones as it ate, and broken bones have been found in its dung. Mei Yong, a *T. rex* discovered in Mongolia, had the nearly complete skeleton of a *Triceratops* (try-SAIR-a-tops) still wedged in its mouth, leading scientists to believe it choked to death. In South America, *T. rex* skeletons have been found with large numbers of small bird fossils in their mouths. It is thought they may have choked on bird feathers.

 FACT **OR** **FICTION**

24

Jobaria (jo-BAR-ee-a) was named after Jobar, a mythical creature of African legends. The herbivore's remains were found in the Sahara Desert in 1997 after being partially uncovered because of erosion. It is thought that *Jobaria* could rear up on its hind legs, like elephants do today. This dinosaur lived about 135 million years ago. From its very long neck to its equally long tail, it was around 21 to 23 m (70 to 75 ft) long and weighed about 18,200 kg (20 US t).

✓ **FACT** **OR** **FICTION**

Freaky Fact or Fiction

25 *Giganotosaurus* (JI-ga-NO-to-SAWR-us), which means 'giant southern lizard', was the longest meat-eating dinosaur. It was probably bigger than *Tyrannosaurus rex* (tie-RAN-uh-SAWR-us rex). It lived more than 90 million years ago and its fossils have been found in Argentina, South America. It walked on two legs and had enormous jaws with long jagged teeth. Its skull was as long as a tall adult human, at 1.8 m (6 ft), but its brain was very small.

 ✓ FACT OR  FICTION

26

In 1878, 39 *Iguanodon* (ig-WAHN-o-don) skeletons were found in a Belgian coal mine. The job of sorting, preparing and describing the collection was given to Louis Dollo. It was not an easy job, as you can imagine. For the first time in Europe, Dollo was able to reconstruct some complete dinosaur skeletons in natural poses. He was also one of the first paleontologists to think about how dinosaurs lived, rather than just giving them a name.

 FACT **OR** **FICTION**

27

Brachiosaurus (BRACK-ee-uh-SAWR-us) was a huge dinosaur, growing up to 15.2 m (50 ft) tall and up to 26 m (85 ft) long. A plant-eater, it had a very long neck made of 14 separate, strong bones. If it stretched its neck up, it would be tall enough to look over a three-storey building of today. It also had very long front legs and a tiny head. A female *Brachiosaurus* usually laid its eggs in lines while walking, rather than in nests. *Brachiosaurus* lived for about 100 years. Skeletons of *Brachiosaurus* have been found in Tanzania, Africa, and in the USA.

✓ FACT   OR   FICTION

28

Some experts say that pieces of rotten meat lodged in its teeth must surely have given *T. rex* a very bad case of halitosis, commonly known as bad breath. On discovery of a large number of dinosaur remains in North Africa in 1991, soil studies revealed that the plant belladonna, also known as deadly nightshade, had once grown in the area. It is suspected that dinosaurs chewed on the sweet plant to get rid of the sour taste in their mouths, causing their deaths.

 FACT **OR** **FICTION**

Freaky Fact or Fiction

29

Imagine a dinosaur that looked like a chicken! The remains of *Gigantoraptor* (ji-GAN-tow-RAP-tor), thought to have weighed 1400 kg (1.5 US t), was discovered by a Chinese professor in Mongolia, an Asian country rich with fossils. *Gigantoraptor* is the biggest bird-like dinosaur. At 5 m (16.4 ft) tall, it is about the size of *Tyrannosaurus rex* (tie-RAN-uh-SAWR-us rex). The 85-million-year-old creature is thought to have had a beak and patches of feathers. The largest known feathered animal before the Chinese discovery was Stirton's Thunderbird, which weighed 0.5 t (0.6 US t) and lived in Australia more than six million years ago.

 ✓ **FACT** **OR** **FICTION**

30

Like all theropods (THERE-uh-pods), *Albertosaurus* (al-BER-tuh-SAWR-us) was a fast, two-legged carnivore with short arms. It had a large head with sharp, slicing teeth and well-developed jaw muscles. It walked on its two strong legs and had bird-like, three-toed, clawed feet. *Albertosaurus* was about 9.1 m (30 ft) long, and about 3.4 m (11 ft) tall at the hips. It weighed about 2500 kg (2.8 US t) and was at one time also known as *Gorgosaurus* (GOR-go-SAWR-us). It is thought to have hunted in packs, and has been found in Asia and North America.

 ✓ **FACT** **OR** **FICTION**

The *Neonasaurus* (NEE-on-a-SAWR-us) was a dinosaur that roamed the Earth some 150 million years ago. It was much smaller than its massive cousins such as *T. rex*. Because of this, it probably lived much of its life in hiding. Studies have shown that it was one of the few burrowing dinosaurs, always staying deep underground at night. If it had stayed above ground, it would have been easily seen by its enemies. The *Neonasaurus* was the only dinosaur whose skin glowed in the dark.

 FACT **OR** **FICTION**

32

Can you imagine a cabin built from fossilised dinosaur bones? That is what confronted paleontologist Walter Granger in 1897 in Wyoming, USA. Built by a shepherd, the small bone cabin was made from fossilised dinosaur bones of the Jurassic period. The area is now known as the Bone Cabin Quarry. It became the site of one of the most important North American finds of Jurassic dinosaur fossils. Nearby, northwest of Laramie, there is another dinosaur building. Constructed in the 1930s, the Fossil Cabin Museum contains fragments from nearly 6000 fossilised bones, some of them 150 million years old.

✓ **FACT**   **OR**   **FICTION**

Freaky Fact
or Fiction

33

The dinosaur *Alamosaurus* (AL-a-moh-SAWR-us) was not named after the Alamo, a famous fort in Texas, USA, as some people believe. Instead, it was named after the Ojo Alamo sandstone formation in New Mexico, USA, where *Alamosaurus* specimens were first found in 1922. Other *Alamosaurus* fossils have been discovered in Utah, Wyoming and the Big Bend region of Texas. This 21.3 m (70 ft) long reptile was the last known sauropod (SAWR-uh-pod) dinosaur, living right at the end of the age of dinosaurs. Some paleontologists think Texas may have been home to 350,000 of these herbivores.

✓ **FACT**　　**OR**　　**FICTION**

34

Late in 2010, scientists published amazing photos of a bird they say is the closest relative to dinosaurs ever found. And it isn't a fossil – it's a living creature! A research team exploring in the Amazon Basin discovered the bird. The size of a toucan and bright green, it has a long whip-like tail, a large crested skullcap and a small, bony horn on the tip of its nose. Scientist Thomas Bull said: 'Tests are still being done, but we are confident that this bird is the missing link. This is the big one!'

 FACT OR FICTION

Freaky Fact or Fiction

35

Ghost Ranch in New Mexico is the home of a colossal fossil bone bed. It was here in 1947 that hundreds of fossilised dinosaur skeletons were found in a jumbled grave. The fossils belonged to a dinosaur called *Coelophysis* (SEE-lo-FIE-sis). A two-legged carnivore, this creature roamed the area over 200 million years ago. It was around 3 m (9.8 ft) long and weighed about 45 kg (100 lb). And it was most likely a cannibal!

 ✓ **FACT** **OR** **FICTION**

36

A dinosaur called *Saichania* (sye-CHAIN-ee-uh) had its own air-conditioning system. Because it lived in a hot climate in Asia, it is thought that its nasal passages gradually adapted to allow it to moisten the air it breathed. Its name means 'beautiful one'. This refers more to the good condition of its recovered bones than its personal beauty. A herbivore, it was about 7 m (23 ft) long. Most of its body bristled with tough plates, some of them spiked. It had a bony club at the end of its tail.

 FACT **OR** **FICTION**

37

It is believed that all tyrannosaurs, including *T. rex*, must have been covered in feathers at some stage of their life, most probably when they first hatched out of their eggs. The largest private collection of dinosaur feathers was owned by Professor Albert Paleo, the father of paleontology, which is the study of prehistoric life. In 1840, he presented Queen Victoria with a cushion made from dinosaur feathers as a wedding gift. Today, the cushion is on display at the British Museum.

 ✓ **FACT**   **OR**   **FICTION**

Dinosaurs

● ●

38

Named after the Aztec feathered-serpent god, *Quetzalcoatlus* (KWET-zal-ko-AT-lus) was a flying reptile. Bones found in the early 1970s in Texas, USA, were bigger that *Pteranodon (TER-an-o-don)* which, until then, was the largest known flying reptile. It had a wingspan of up to 15 m (49 ft) and was able to fly long distances. It had an extremely long neck, slender jaws with no teeth and a long, bony crest on its head. Some scientists believe *Quetzacoatlus* was like a vulture and fed on the bodies of dead dinosaurs.

 FACT   **OR**   **FICTION**

Freaky Fact or Fiction

39

Sometimes you don't have to go out into the wilderness to find lost dinosaurs. They might be right under your nose, in a museum. This was the case when Englishman Mike Taylor visited London's Natural History Museum in 2006. A PhD student who had been studying sauropod (SAWR-uh-pod) vertebrae, he noticed a dinosaur bone that he had never seen before. Incredibly, it was later found to be a completely new species. It is called *Xenoposeidon* (ZEE-no-puh-SY-don), which means 'alien sauropod'. It had been lying unnoticed in a basement of the museum for more than 100 years.

 ✓ **FACT** **OR** **FICTION**

Dinosaurs

When the dinosaurs lived, there were many aquatic reptiles in the seas, such as plesiosaurs (PLEE-see-uh-sawrs), nothosaurs (NOTH-o-sawrs), mosasaurs (MO-suh-sawrs) and ichthyosaurs (IK-thee-uh-sawrs). However, these water reptiles were not dinosaurs. Another incorrect idea that people have is that hairy mammoths and sabre-toothed tigers or *Smilodon* (SMY-lo-don) lived at the same time as dinosaurs. In fact, they lived millions of years later. The mammoth became extinct 10,000 years ago and it is believed that some sabre-toothed tigers were alive just 4000 years ago!

✓ **FACT**   **OR**   **FICTION**

Freaky Fact or Fiction

41 It is difficult to tell if dinosaurs were smart. Usually intelligence is related to a large brain in a small body. The Cretaceous bird-like dinosaur *Troödon* (TRO-uh-don) may have been intelligent for this reason. It was a fast mover and its large eyes were equipped with stereo vision. *Troödon*'s brain was relatively large and it is often listed as the smartest dinosaur. Two others that may have been a little brighter than the rest are *Deinonchysus* (dye-NON-i-kus) and *Compsognathus* (komp-sog-NAY-thus).

✓ FACT   OR   FICTION

42

Dinosaurs became extinct millions of years before the first humans lived on Earth. When Neanderthal Man did at last appear, about 120,000 years ago, dinosaur bones were scattered everywhere. Having no other materials on hand, early man sharpened these bones by rubbing them together, and then used them in the tips of arrows and spears. As well, it is believed that the first necklaces and bracelets were made from dinosaur bones.

✓ **FACT**   **OR**   **FICTION**

Freaky Fact or Fiction

43

If you camp out in the great outdoors, you will most likely wake up to hear birds singing their morning songs. Scientists believe that dinosaurs also communicated with each other using something that might have been similar to birdsong. Many dinosaurs had complex crests that could produce low rumbles, which in some cases might have sounded like a foghorn – and we all know singers who sound like that. Another communication method might have been a mighty crack of a dinosaur tail. And finally, just as today's birds dance for their mates, it is probable that some dinosaurs danced.

 ✓ **FACT**   **OR**   **FICTION**

The arrival of *Plateosaurus* (PLAY-tee-uh-SAWR-us) marked a change in the development of dinosaurs. Before it appeared over 200 million years ago, herbivores had short necks. This made it impossible for them to reach food that was found high up in trees. Then along came long-necked *Plateosaurus*. Paleontologists believe it also had the ability to stand on its hind legs to reach trees for their food. It grew to around 9 m (29.5 ft) long and up to 4 m (13.1 ft) tall. Its name means 'flat lizard'.

 ✓ **FACT** **OR** **FICTION**

45

In 2009, three new species of dinosaur were found in the Winton area of Queensland, Australia. The dinosaurs were given the nicknames Banjo, Matilda and Clancy. The poet Banjo Paterson is said to have written the classic Australian poem 'Waltzing Matilda' while at Winton in 1885, and he also wrote the famous poem 'Clancy of the Overflow'. Banjo is similar to *Velociraptor* (vel-os-i-RAP-tor), but is smaller. Matilda is similar to the hippopotamus of today, and Clancy is a tall animal that may have been Australia's prehistoric equivalent of the giraffe.

✓ **FACT**   **OR**   **FICTION**

● ● ● ● ● ● ● ● ● ● ● ● ● ● ● ● ● ● ● ●

46

In 1982, Ernst and Bobby-Sue Furphy were on a camping trip in Tupelo, Mississippi, when they accidentally made an important scientific discovery. While exploring caves in the area, they found a large mass of very old and very large bones. A research team soon identified the remains as being from a previously unknown dinosaur species. Because Tupelo is the birthplace of famous singer Elvis Presley, the Furphys named it *Presasaurus* (PRES-uh-SAWR-us).

 ✓ **FACT** **OR** **FICTION**

Freaky Fact or Fiction

Like reptiles and birds today, some dinosaurs laid hard-shelled eggs. We know this because many fossilised dinosaur eggs have been found. Some of them were in nests. The first evidence of this was in the 1920s in Mongolia when a nest of eggs laid by a *Protoceratops* (pro-tow-SAIR-a-tops) was discovered. This showed scientists that baby dinosaurs, like birds, probably stayed in their nests. Several nests have been found close together. This suggests that some dinosaurs

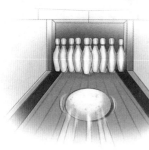 nested in colonies. In 1869, the first fossilised dinosaur eggs were found in France. These oval-shaped eggs are as heavy as a bowling ball.

 ✓ **FACT** **OR** **FICTION**

Dinosaurs

Not much is known about the *Liliensternus* (LIL-ee-en-STERN-us), a dinosaur from the Late Triassic period. This is because there have been only two incomplete skeletons found. It was first found in 1934 in Germany and named after a German scientist. *Liliensternus* had a crest on its head. It was up to 5 m (16.5 ft) long, and may have weighed about 127 kg (280 lb). Scientists think it was a carnivore and might have preyed on plant-eaters.

✓ **FACT** **OR** **FICTION**

49

Imagine a terrifying monster – a 12 m (39.4 ft) long lizard – is about to eat you! If you lived in North America in the Cretaceous period, you might come face to face with the meat-eater *Acrocanthosaurus* (Ak-ro-KANTH-uh-SAWR-us). It had large spines on its backbone that it probably raised in a kind of sail. This might have been like the colourful feathers of a bird – used to warn other dinosaurs to stay away. Or it might have been used to attract a mate. *Acrocanthosaurus* fossils have been found in North America.

✓ **FACT** **OR** **FICTION**

50

Sometimes there is debate about whether a creature was really a dinosaur. *Dimorphodon* (dy-MORE-fo-don) is one such example. It might easily be known as 'big head', because its head was huge. It was even bigger than its body! Its neck was long and its rib cage very small. Like a kangaroo, it seems to have used its long tail to help it balance as it ran along on its two legs. This creature was actually a flying reptile, not a dinosaur as many people think. It had wings made from skin. It had two types of teeth in its jaws, which is rare for flying reptiles. Its main diet is thought to be fish.

✓ FACT   OR   FICTION

51

Dollodon (DOLL-uh-don) was a two-legged meat-eater that lived 145 million years ago. Its fossil remains were found in Transylvania, the European home of Dracula. The Transylvanians were so proud of finding the dinosaur's remains they created a toy. This toy, which was a replica of how scientists believed *Dollodon* looked, was very popular.

✓ **FACT**   **OR**   **FICTION**

52

There are many ways to choose a dinosaur name. Usually their names come from Latin or Greek words. Some are named after the place where they are found, such as *Argentinosaurus* (AHR-gen-TEEN-uh-SAWR-us). There are those named because of their size, such as *Megalosaurus* (MEG-uh-lo-SAWR-us) – very big – or *Bambiraptor* (BAM-bee-RAP-tor) – very small. Others are named to honour the person who discovered them, and a smaller number are named after famous people. Others still are named from what we know of their behaviour. For instance, *Velociraptor* (vel-os-i-RAP-tor) means 'speedy robber'.

✓ **FACT**   **OR**   **FICTION**

53

Fossil remains of *Chirostenotes* (kie-ROSS-ten-oh-tease) have been found in Canada. This dinosaur was found in sections. First, its long, narrow hands were found in 1924. This gave it the Greek name for 'narrow hand'. Its feet were found a few years later, and its jaw was discovered a few years after that. Imagine what a job it must be for a scientist to piece all of the parts together! *Chirostenotes* had a beak, long arms with strong claws, long, thin toes and a high, rounded crest on its head – like a modern-day cassowary. Most likely it had feathers.

✓ **FACT**   **OR**   **FICTION**

● ● ● ● ● ● ● ● ● ● ● ● ● ● ● ● ● ●

54

Flying dinosaurs! In outer space! It sounds like a very bad Hollywood movie, but it's a fact. Dinosaurs have flown into space! The first dinosaurs in space were *Maiasaura* (may-ya-SAWR-a) and *Coelophysis* (SEE-lo-FIE-sis). In 1985, astronaut Loren Acton took a piece of *Maiasaura* bone and a piece of *Maiasaura* eggshell on an eight-day Spacelab-2 mission. In 1998, a *Coelophysis* skull was taken into space by the space shuttle *Endeavor,* which travelled to the space station *Mir.*

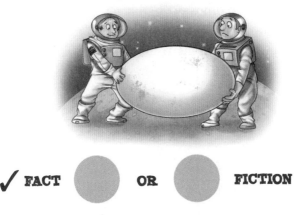

✓ **FACT**   **OR**   **FICTION**

55

Fossil remains of *Atlascopcosaurus* (AT-lus-KOP-kuh-SAWR-us) were found at Dinosaur Cove in Victoria, Australia, in the 1980s. It was a herbivore that was about 2 to 3 m (6.6 to 9.8 ft) long and was probably a fast runner. *Atlascopcosaurus* got its name from the Atlas Copco Company. This company supplied equipment for the expedition that discovered the dinosaur. Examination of a partial skeleton believed to be from an *Atlascopcosaurus* showed that in its final years the dinosaur had a severe bone infection in its tibia (shin bone), but still managed to survive for a long time while it was ill.

✓ FACT   OR   FICTION

An African inventor, Sh'ana Bernard, discovered fossil bones while searching for firewood in Niger in 1962. She was planning to use the bones in an invention that helped draw water from a well. However, she first showed the bones to a visiting paleontologist. Alfred Leakey identified the bones as that of a dinosaur. The new discovery was named *Afrovenator* (AF-roh-VEN-uh-tehr) in honour of Ms Bernard.

 ✓ FACT OR FICTION

57

Anchiornis (AN-key-OR-nis) was a dinosaur that was only about the size of a chicken. What makes it big in the dinosaur world is the fact that it had feathers. Not only that, but scientists also now know what colour it was. While other dinosaurs had crests of bony plates and spikes, *Anchiornis* had a reddish-brown feathered crest, in a Mohawk shape. It also had a pattern of black and white stripes on its wings and feet. Its name means 'almost bird', but it is definitely a dinosaur.

✓ **FACT**   **OR**   **FICTION**

58

Clams were most likely the favourite food of *Dsungaripterus* (JUNG-gah-RIP-ter-us) from China and South America. This flying reptile had a wingspan no larger than 3 m (9.8 ft). It had an upturned upper jaw, which probably helped it snap open clams. It had no teeth at the front of its mouth, but the back contained broad, blunt teeth, which may have helped in crushing the shells. It probably also ate fish, other molluscs, crabs, insects and dead land animals. Scientists believe *Dsungaripterus* lived in colonies.

✓ **FACT**   **OR**   **FICTION**

Freaky Fact or Fiction

59 *Edmontosaurus* (ed-MON-tuh-SAWR-us) was a hadrosaur (HAD-ruh-SAWR), or duck-billed dinosaur. A herbivore, it weighed up to 3175 kg (3.5 US t) and was about 12.8 m (42 ft) long. Despite it size, scientists have discovered bite marks from *T. rex* on at least one *Edmontosaurus*. It was not armoured, so had little defence against bigger animals. *Edmontosaurus* had as many as 1000 teeth, but these were tiny and suited to munching vegetation, not snapping at a *T. rex*. Named for the area in Canada where it was found, *Edmontosaurus* is one of the last known dinosaurs to have lived.

✓ FACT   OR   FICTION

60

Chunkingasaurus (Chunk-ING-ah-SAWR-us) was so named because frozen chunks of the dinosaur were found in Antarctica. This discovery in 1932 was remarkable because all previous dinosaur remains found were fossilised bones. *Chunkingasaurus*'s meat was very well preserved. Among the remains was a stomach that contained undigested penguins, polar bears and walruses. The dinosaur body is now on display in the Natural Museum of Auckland, New Zealand.

 ✓ **FACT** **OR** **FICTION**

61

Gargoyleosaurus (gar-GOYL-e-oh-SAWR-us) means 'gargoyle lizard'. *Gargoyleosaurus* lived during the Late Jurassic period, about 154 to 144 million years ago, and was a herbivore. Paleontologists think that *Gargoyleosaurus* was the first true ankylosaur (an-KIE-luh-SAWR), and the smallest. These armoured dinosaurs were short and kept low to the ground. The idea of this was probably to make predators think the ankylosaur would be difficult to eat. *Gargoyleosaurus* walked on four stubby legs and was about 3 to 4 m (9.8 to 13.1 ft) long. Fossils were found in North America.

✓ **FACT**   OR   **FICTION**

● ●

62

K*entrosaurus* (KEN-truh-SAWR-us) always wore a full suit of armour. From its neck, almost to the point of its tail, it had large pointed spikes. More spikes protruded from its hips, and it had bony plates on its shoulders. No wonder its name means 'prickly lizard'. Discovered in eastern Africa, *Kentrosaurus* was less than 2 m (6.6 ft) tall and 3.5 (11.5 ft) long. A slow-moving herbivore, it was probably attacked by larger dinosaurs.

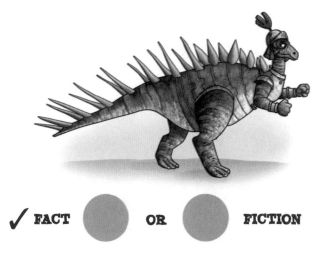

✓ **FACT**   **OR**   **FICTION**

63

Herbivorous dinosaurs are usually notable for their size. They were giants. But *Pisanosaurus* (pe-ZAHN-uh-SAWR-us) was the exception to the rule. This herbivore was only about 1 m (3.3 ft) long and 30 cm (11.8 in) tall. Not a lot is known about it because only small fragments of its remains have been found. We do know that it is one of the oldest dinosaurs, dating back to more than 200 million years ago. It was named to honour paleontologist Juan A Pisano.

 ✓ **FACT** **OR** **FICTION**

64

The *Telmatosaurus* (tel-MAT-oh-SAWR-us) is thought to have lived shortly before the mass extinction of the dinosaurs. One of the few dinosaurs belonging to the duck-billed dinosaur group known to have lived in Europe, its name means 'swamp lizard'. Walking on two legs, it was a plant-eater. Its height was 2 m (6.6 ft) and it was 5 m (16.4 ft) long.

✓ FACT   OR   FICTION

65 Ichthyosaurs (Ick-thee-oh-saws), meaning 'fish-lizards', were some of the largest marine animals, growing up to 10 m (about 33 ft) long. The earliest ichthyosaurs lived in the Late Triassic period, more than 200 million years ago. They were similar in appearance to the modern-day dolphin, though they were reptiles, not mammals or fish. Even though they were reptiles, ichthyosaurs gave birth to live young rather than laying eggs.

 ✓ FACT   OR   FICTION

• •

66

Scientific studies have shown that birds are descendants of the group of dinosaurs known as theropods (THERE-uh-pods). Some of the largest dinosaurs, including *T. rex*, were theropods. However, birds came from a different, much smaller branch of the group. It is argued that since birds and dinosaurs are very distantly related, while birds still exist, dinosaurs are not really extinct. Theropods were flesh-eating dinosaurs who used their hind legs for support and movement. Their short upper limbs were probably used for grasping and tearing food. Theropod feet usually looked like those of birds.

 ✓ **FACT**　　**OR**　　**FICTION**

Freaky Fact or Fiction

67 Have you ever wondered how dinosaur fossils are dug up? It's a job that is done very slowly and carefully. Dinosaur bones are chipped out from rock with chisels, drills and hammers. Loose rock and sand is brushed away. When the bones are ready to go to a museum, they are first sprayed with a glue to keep them strong. Then they are wrapped in bandages soaked in plaster of Paris. Or they are put into a parcel of polyurethane foam. This protects them from damage.

 ✓ FACT   OR   FICTION

68

The grounds of the White House in the US may hold dinosaur treasure. While patrolling the lawn area in 2010, a security guard noticed the president's dog behaving strangely. On closer inspection, he found the dog was trying to pull a bone from the ground. Tests have revealed the bone is that of a *Bambiraptor* (BAM-bee-RAP-tor). Researchers say that there may be as many as 100 dinosaurs buried near the White House. Because of this, Congress is considering whether the historic building should be demolished and rebuilt elsewhere. The President has stated that he does not wish to stand in the way of valuable research.

 ✓ **FACT** **OR** **FICTION**

69

Zuniceratops (ZOO-nee-SAIR-a-tops), the oldest horned dinosaur ever found, was discovered in New Mexico in 1996 by Christopher James Wolfe, aged eight. Christopher was with his father, a paleontologist, when he found a piece of the dinosaur's horn. The complete name of the species is *Zuniceratops christopheri* (ZOO-nee-SAIR-a-tops kris-TOFF-e-ree) – it was named in honour of its discoverer.

✓ **FACT** **OR** **FICTION**

70

A newly discovered dinosaur species – *Dracorex hogwartsia* (DRAY-koh-rex hog-WART-see-uh) – is named in honour of author JK Rowling and her Harry Potter books. The dragon-like dinosaur's name comes from the Latin word *draco* meaning 'dragon', *rex* meaning 'king', and *hogwartsia*, which stems from Rowling's fictional Hogwarts School of Witchcraft and Wizardry. Science fiction writer Arthur C Clarke has also had a dinosaur named after him, as has Michael Crichton, author of the book *Jurassic Park*.

 ✓ **FACT** **OR** **FICTION**

Freaky Fact or Fiction

71 Have you ever been to a museum and seen the re-creation of a dinosaur? Museum workers wire individual fossilised dinosaur bones together and put the whole skeleton on a metal frame for display. If bones are missing, they make false bones of plaster or plastic. From marks on bones where muscles were attached, scientists can tell which bones belong where, and also how the dinosaur moved.

 FACT   OR   **FICTION**

72

Baryonyx (bar-ee-ON-iks) looked like a large, menacing meat-eater but studies of its skull show that it was more like a fish-eating crocodile. A spinosaur (SPY-nuh-SAWR), *Baryonyx* had 30-cm-long (1-ft-long) front claws, which it may have used for catching fish, much like modern-day bears catch salmon. It was 9 m (30 ft) long and lived about 125 million years ago. Spinosaurs were a family of fish-eating dinosaurs, some of which were bigger than *T. rex*.

✓ **FACT**   **OR**   **FICTION**

Freaky Fact
or Fiction

73

efore the first dinosaurs, there were mammal-like reptiles. But when dinosaurs came, they hunted and ate the much smaller, furry animals. However, as the dinosaur population decreased, the early mammals began to multiply. When the dinosaurs disappeared, the mammals took over. Within a few million years, some of the early mammals evolved into the earliest types of horses, elephants and camels.

 ✓ FACT **OR** **FICTION**

Dinosaurs

After many years of research, scientists at Princedom University have concluded that dinosaurs were very itchy. It seems wherever dinosaurs lived there were always huge boulders for them to scratch themselves against. Scientists have pinpointed more than 50 of these boulders, which are still intact today. Some are more than 100 million years old. Soil samples taken from near the boulders have revealed dinosaur DNA. There are also traces of dinosaur hair and fur, as well as dandruff.

The most exciting find comes from Papua New Guinea. There, scientists have discovered microscopic evidence of dinosaur fleas.

 FACT **OR** **FICTION**

Freaky Fact or Fiction

75 For decades it was commonly thought that dinosaurs such as *Pentaceratops* (PEN-tah-SAIR-a-tops) and *Triceratops* (try-SAIR-a-tops) used their sharp horns and spikes to fight off predators. This may be true to some extent, but it is not their main purpose. It is now thought that the horns and spikes allowed them to recognise their own species, and to compete for mates by locking horns, much as horned animals still do today. The horns might also have been used to attract females.

✓ FACT   OR   FICTION

● ● ● ● ● ● ● ● ● ● ● ● ● ● ● ● ● ● ●

76

The fossil *Archaeopteryx* (ahr-kee-OP-ter-iks) is well known as the 'missing link' between reptiles and birds. It had feathers like a bird but its skeleton was similar to that of a small, meat-eating dinosaur. *Archaeopteryx* had feathers on its wings and tail. It could probably glide, but may not have been able to fly by flapping its wings.

✓ **FACT** **OR** **FICTION**

77

For hundreds of years adventurers have searched for the *Doradosaurus* (dor-AH-do-SAWR-us). This name comes from the Spanish words *El Dorado*, meaning 'the golden one'. It was so named because its tail bone held heavy deposits of gold. The last major discovery of *Doradosaurus* bones was at Canada's Yukon River in 1896. A massive gold rush was triggered when news spread that the Yukon and nearby Klondike contained *Doradosaurus* graveyards. More than 100,000 people rushed there, picking the animals clean. All traces of the beast disappeared after that, but to this day scientists are still searching for the mysterious *Doradosaurus*.

 FACT OR **FICTION**

Some dinosaur bones are collected on expeditions by specialists. But amateur collectors can make great discoveries. In 1983, Bill Walker made a remarkable find – a huge curved claw. It was in a clay pit in southern England. This claw showed that it must have come from a giant carnivore that had never been found anywhere else in the world. The skull of the dinosaur was long and flat like that of a crocodile. *Baryonyx walkeri* (bar-ee-ON-iks WALL-ker-eye) was named after Mr Walker.

 FACT **OR** **FICTION**

Freaky Fact or Fiction

79 Horned dinosaurs are called ceratopsian (sair-a-TOP-see-an), or horned-face dinosaurs. The biggest and most famous ceratopsian is *Triceratops* (try-SAIR-a-tops). In 2002, the remains of a dog-sized ceratopsian was discovered in China. Named *Liaoceratops* (lee-OW-SAIR-a-tops), it has two small horns, one below each eye. At 130 million years old, this creature is one of the oldest and smallest horned dinosaurs.

 FACT   **OR**   **FICTION**

Most of us have seen movies, such as *Jurassic Park*, in which a ferocious *T. rex* gallops alongside a fast-moving car, terrifying the passengers. Recent studies have shown that this could not have been true. *T. rex* weighed in at over 6000 kg (13,228 lb). The dinosaur would have needed most of its body weight in its legs to be able to run as fast as a car going at 72 km/h (45 mi/h). Most likely it moved slowly, so that a fast runner or a person on a bike could have escaped.

✓ **FACT**   **OR**   **FICTION**

Freaky Fact or Fiction

81

Battles were fought over dinosaur bones. The most famous 'Bone War' took place in the United States. Two men, Edward Drinker Cope and Othniel Charles Marsh, both fossil hunters, competed against one another to find the most dinosaur bones. They fought each other and also had to fend off hostile Native Americans, who were defending their lands. This took place in the 1870s and '80s. Between them, Cope and Marsh discovered 136 new kinds of dinosaurs.

 FACT OR FICTION

A dino-snore sounds like a joke, but it really happened! Like humans, dinosaurs had soft tissue at the back of their throats. The soft tissue would silently vibrate when the dinosaurs breathed in and out. However, for some larger dinosaurs such as sauropods (SAWR-uh-pods), their giant tongues would interfere with the vibrations. When the dinosaur was sleeping, its tongue, which measured up to 1 m (3.28 ft) in length and weighed up to 150 kg (330 lb), would relax and press against the vibrating soft tissue at the back of the throat. This would cause an almighty sound . . . a dino-snore!

✓ FACT OR FICTION

83

Australian researchers have found evidence of what could be the world's only surfing dinosaurs. Excavations at the famed Bondi Beach in New South Wales have uncovered the fossilised remains of a new dinosaur that has been named *Dudeasaurus* (DOOD-uh-SAWR-us). This dinosaur stood on two thin legs, and had very large, flat feet – almost like miniature surfboards. *Dudeasaurus* was a strong swimmer and is thought to have 'surfed' regularly in order to catch fish. There is also a theory that the male dinosaurs may have used their surfing ability to impress female dinosaurs.

✓ **FACT**   OR   **FICTION**

84 At Danny and Dawn's Dinosaur Restaurant in Wollongong, Australia, there is a distinct dinosaur flavour. It starts with the Jurassic Parking Lot and continues with the menu. On it you will find freshly caught *T. rex* fillets, baked *Brontosaurus* (BRON-toe-SAWR-us) beans, eye of *Iguanodon* (ig-WAHN-o-don) soup and poached pterodactyl (TER-uh-DAK-til) wings. There is also a warning on the menu: Be careful! All our food has bones!

 FACT **OR** **FICTION**

85

New evidence shows that Australia had its own tyrannosaur (tie-RAN-uh-SAWR), but it was a much smaller animal than its famous and ferocious relative *T. rex*. A piece of fossilised pelvic bone was actually found at Victoria's Dinosaur Cove in 1989, but was not identified as tyrannosauroid until 20 years later. Scientists say the bone once belonged to a miniature tyrannosaur with small arms and powerful jaws; it was only about 3 m (9.8 ft) long and weighed 80 kg (176 lb). This is the first evidence that the tyrannosaur family existed in Australia.

 ✓ FACT   OR   FICTION

86

Fossil remains have been found in India of a large snake coiled around a broken egg in a dinosaur nest. It seems the snake was about to eat a baby titanosaur (tie-TAN-uh-SAWR), when a storm or landslide hit, burying both the snake and dinosaur under a mountain of sand and mud. There they lay for 67 million years before being discovered. Other snake skeletons and dinosaur eggs were also found at the site. It is the first evidence that snakes ate dinosaurs.

 ✓ FACT **OR** **FICTION**

87 Scientists have now confirmed that the turkey-sized *Sinosauropteryx* (sye-nuh-SAWR-OP-ter-iks), which roamed the land of modern-day China about 125 million years ago, was covered in orange and white feathers, and had white stripes on its tail. Rather than using them for flying, camouflage or warmth, the feathers were most probably for display purposes. Professor Michael Benton from the University of Bristol said that feathered dinosaurs might have come in many colours.

 ✓ **FACT** **OR** **FICTION**

38

Scientists have found that dinosaurs were the big winners when volcanic eruptions shook the world 200 million years ago. Paleontologists say the eruptions killed off the dinosaur's main rivals, the crurotarsans (crew-ro-TAR-sans). These creatures were closely related to modern-day crocodiles. Scientists still don't know why the dinosaurs survived the eruptions while the crurotarsans were destroyed.

✓ **FACT**   **OR**   **FICTION**

Freaky Fact or Fiction

89 *The Chambridge Dictionary* has recently added several new words to its list, and they all relate to dinosaurs. A baby dinosaur, whether male or female, is known as a cubasaur (CUB-uh-SAWR). A mature male dinosaur is a roarasaur (RAW-uh-SAWR), while a mature female is a msasaur (MIZZ-uh-SAWR). A vicious dinosaur is a snapasaur (SNAP-uh-SAWR). And a timid dinosaur is a wussasaur (WOOSS-uh-SAWR). Finally, the collective name for a herd of dinosaurs is a thump.

 FACT **OR** **FICTION**

The fastest dinosaur was probably *Struthiomimus* (strooth-ee-uh-MY-mus). This animal was shaped like an ostrich, but had a long tail. Its name actually means 'ostrich mimic'. It is estimated that *Struthiomimus* could have run as fast as an ostrich or a horse. It needed to be fast because bigger dinosaurs liked to eat it. It had no teeth, just a horny beak, and some scientists think it was mainly a herbivore.

 ✓ **FACT** **OR** **FICTION**

91

Iguanodon (ig-WAHN-o-don) was a herbivorous dinosaur that lived in Europe, North America, North Africa, Australia and Asia, mostly during the Late Jurassic and Early Cretaceous periods between about 161 million and 100 million years ago. *Iguanodon* weighed as much as 4.5 t (nearly 5 US t), were about 9 m (30 ft) long, and stood up to 9 m (30 ft) tall on their hind legs. *Iguanodon*'s long, flat head ended in a horny beak, and its jaws contained teeth that looked like those of the modern-day iguana lizard. Its name means 'iguana tooth'.

 ✓ FACT   **OR**   **FICTION**

British fossil hunter and clergyman William Buckland (1784–1856) was the first person to scientifically describe and name a dinosaur. In 1824, he gave the name *Megalosaurus* (MEG-uh-lo-SAWR-us) to a dinosaur whose fossil remains had been found nearly 200 years before. Buckland always collected his fossils in a large blue bag, which he nearly always carried with him. At the time he named *Megalosaurus*, the word dinosaur hadn't even been invented yet.

 FACT OR  **FICTION**

93

Megalosaurus (MEG-uh-lo-SAWR-us), which means 'great lizard', walked on two strong legs and had a massive tail. It had short arms and three-fingered hands with sharp claws. It was up to 9 m (29.5 ft) long, 3 m (9.8 feet) tall and weighed about 1 t (1.1 US t). Like *T. rex* and other large dinosaurs, it may have been a scavenger as well as a hunter. When some of its large bones were first discovered, it was thought they belonged to a giant man.

✓ **FACT**   **OR**   **FICTION**

94

Field tests at Hollywood University have shown that North American dinosaurs were migratory animals. It seems dinosaurs usually coped fairly well with the chilly temperatures of places such as present-day New York or Boston. However, the winter cold was too much for them. When the snows came many thousands of them would make

their way to Miami and Florida, where it was much warmer. Las Vegas was also a popular destination. Scientists are almost certain this great migration began on the same day every year. This caused the world's first traffic jam.

 FACT **OR** **FICTION**

95

Boston Professor Horace Rubble hotly disputes the theory that dinosaurs became extinct after a huge asteroid crashed into Earth. 'The real reason is that dinosaurs died off because they were bored,' Professor Rubble said. 'It is so obvious. They lived in a world where there was nothing to do but eat – and what boring foods they had! Believe me, if an asteroid had crashed into Earth the dinosaurs would have been happy. It would have given them some entertainment!'

 FACT **OR** **FICTION**

96

When dinosaur bones were first found, they were thought to be the bones of giants. The first remains that were recognised as those of huge reptiles were supposedly found by Mary Mantell in 1822. Mary apparently found fossilised teeth in Sussex, England. Her husband, Dr Gideon Mantell, studied the fossilised teeth and some bones, and decided they were like those of the iguana lizard, but much bigger. He later named the big reptile *Iguanodon* (ig-WAHN-o-don).

 ✓ **FACT**   **OR**   **FICTION**

97 A 150-million-year-old landing strip for pterosaurs (TER-uh-SAWRS) was found in France in 2009. The 'runway' shows that the reptiles landed feet first, then staggered before walking on all fours. However, scientists still do not know how the pterosaurs took off. The tracks were made by small pterosaurs, which had a wingspan of 1 m (3.3 ft) and feet that were only 5 cm (2 in) long.

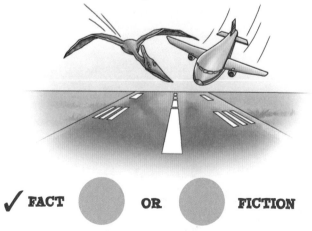

✓ **FACT**   **OR**   **FICTION**

98

Allosaurus (AL-uh-SAWR-us) was a large, carnivorous dinosaur that lived around 150 million years ago, during the Late Jurassic period. These dinosaurs reached 12 m (39.4 ft) long, stood more than 4.5 m (14.8 ft) tall and weighed up to 3.6 t (nearly 4 US t). *Allosaurus* was a biped; it walked on two stout hind limbs and had large bird-like feet, using its heavy tail for balance. It had sharp, grasping claws on its toes and on the hands of its short forelimbs. *Allosaurus* was probably a scavenger as well as a hunter, and may have hunted in groups.

✓ FACT   OR   FICTION

99

Titanosaur (tie-TAN-uh-SAWR) bones have been found on every continent in the world. Scientists have worked out that titanosaurs were giant herbivores that walked on all fours and weighed up to 100 t (110 US t). Most titanosaurs were walking fortresses. Their armour consisted of solid bony plates that covered their bodies and protected them. It is thought that titanosaurs were herd animals, often travelling in large packs. The reason why animals herd is because it offers them some protection against predators. You may have heard of the saying: 'There's safety in numbers'.

✓ FACT　　　OR　　　FICTION

Dinosaurs

A research team led by Professor Ian 'Edgy' Border of Camberbridge has found that dinosaur bones may prove valuable to gardeners. Microscopic analysis of *Triceratops* (try-SAIR-a-tops) bones has detected huge deposits of odium, an extremely rare fertiliser. It is estimated that the crushed bones of just one large dinosaur would provide enough odium for every garden in England. Professor Border has called for more tests to be done before using dinosaur odium as a fertiliser. 'We know it would give us giant vegetables,' he said. 'But we need to make sure it doesn't give us giant people, too.'

✓ **FACT**   **OR**   **FICTION**

Freaky Fact or Fiction

101 There were flying reptiles during the dinosaur era, but did birds evolve from dinosaurs? Some people believe that birds evolved from a group of reptiles called thecodonts (THEE-cuh-donts) during the Triassic period (248 to 208 million years ago.) The problem is that no good fossils of birds have been found dating back beyond about 160 million years ago, let alone to Triassic times.

✓ FACT   OR   FICTION

02

Coelurus (see-LOOR-us) had hollow bones and its name means 'hollow tail'. Its skull would have fit into a human hand and it was as tall as a man. It was lightly built for speed and action, and walked on its two back legs. *Coelurus* ate small vertebrate animals, such as lizards and early mammals. It lived in the forests and swamps of the Late Jurassic period.

 ✓ **FACT** **OR** **FICTION**

103 An almost complete skeleton of *Compsognathus* (komp-sog-NAY-thus) was found in 1972 in France. At first it was thought to have had flippers on its front feet – to help it swim. This was an amazing 'discovery'. However, further studies proved this was not so. There have only been two *Compsognathus* skeletons found. The other was in Germany, in the 1850s. Both were the size of chickens. Its name means 'pretty jaw'.

 ✓ **FACT**   **OR**   **FICTION**

04

Texas millionaire Bubba 'T-Bones' Jones has been sentenced to ten years jail for 'dinosaur stealing'. One day in 1990, the unemployed Jones was walking through a field when he noticed a large bone jutting up from the ground. It was a dinosaur bone! Jones quickly discovered that the whole field was littered with ancient bones. For the next 18 years, he charged people US$1000 (A$1180 or £690) a day to dig in the field. He was arrested in 2009 when it was found that he did not own the land. By that time his fortune had grown to US$13 million (A$15.3 million or £9 million)!

✓ **FACT** **OR** **FICTION**

105

People have sometimes made the wrong assumption that people and dinosaurs lived at the same time. This could be because of some films that showed this happening. One such film was *One Million Years BC* (1966). Another was *The Lost World* (1960), which sent explorers to a weird place in the South American Amazon rainforest where dinosaurs had survived. In *The Last Dinosaur*, made in 1977, some oil prospectors in Antarctica defrost a *Tyrannosaurus rex* (tie-RAN-uh-SAWR-us rex). Mention must be made, too, of *The Flintstones*, a movie and television show in which a prehistoric family had a pet dinosaur named Dino.

✓ **FACT**   **OR**   **FICTION**

06

Insects called dung beetles help droppings from large animals, such as elephants or horses, break down into soil. It is likely that dung beetles were around during the dinosaur years. They had a lot of hard work to do since one dropping from a dinosaur such as a *Stegosaurus* (STEG-uh-SAWR-us) would fill a large household garbage bin!

✓ **FACT**   OR   **FICTION**

Freaky Fact or Fiction

107 Until recently, scientists believed grass evolved after the dinosaurs. But there is now proof that some dinosaurs did eat grass, because grass remains have been found in their stomachs, and fossilised dung. However, the grass may have been really tall – perhaps as tall as a man!

✓ **FACT** **OR** **FICTION**

● ●

08

The *Blottosaurus* (BLOT-uh-SAWR-us) was unique to a small section of the Gobi Desert. Similar in height and shape to the giraffe, this long-necked dinosaur lived entirely on jungle juice berries found high up in sauropod (SAWR-uh-pod) trees. The berries contained a small amount of alcohol. Since the *Blottosaurus* ate them day and night, scientists believe they may have often been intoxicated. It is suspected that the sauropod trees could have been destroyed by the *Blottosaurus* during a drunken rampage, leaving them nothing to eat. If so, this means the *Blottosaurus* is the only dinosaur that wiped itself out.

✓ **FACT**　　　**OR**　　　**FICTION**

109

The first dinosaurs lived in a very different environment to the modern world. At the time of the first dinosaurs there were no flowering plants. There were giant forests with trees such as cycads, conifers, ferns and gingkoes. The first flowering plants, such as passion flowers and magnolias, appeared in the Cretaceous period. Other flowers such as buttercups came later. In 1989, one of the oldest known flowers was discovered in 115- to 118-million-year-old rocks at Koonwarra in South Gippsland, Australia.

✓ FACT　　OR　　FICTION

Most people know that flamingos get their pink-coloured feathers from eating crustaceans (such as prawns or shrimp). Millions of years before flamingos existed, *Pterodaustro* (ter-o-DAWS-tro) had very bristly teeth in its bottom jaw. It may have used these like a sieve, to filter out small animals such as crustaceans from mouthfuls of water. Perhaps this flying reptile was pink like a flamingo!

✓ **FACT** **OR** **FICTION**

Freaky Fact or Fiction

111

In 2006, American paleontologist Anthony Martin found the first evidence that dinosaurs burrowed in the earth. While working in Montana, USA, Martin discovered three small dinosaurs in an ancient fossilised burrow. A year later he was in Australia. He had gone to an isolated area called Knowledge Creek, in Victoria, to search for dinosaur tracks. Yes, he did find dinosaur tracks that day. Incredibly, he also found three more dinosaur burrows. The Australian find is said to be about 106 million years old, making it 11 million years older than the US discovery.

✓ FACT   OR   FICTION

● ●

12

It had long been thought that the first feathered dinosaurs lived some 150 million years ago. This changed in 2009 when the discovery of *Tianyulong confuciusi* (tee-AN-yu-long con-FUCE-us-ee) was announced. A small herbivore about the size of a cat, this feathered dinosaur lived about 198 million years ago. This means that it, and perhaps other feathered dinosaurs, might have lived when dinosaurs first appeared on Earth. The fossil remains of *Tianyulong confuciusi* were discovered by a team of Chinese scientists. Now scientists are excited about the possibility of finding other 'feathered friends'.

✓ **FACT**　　　**OR**　　　**FICTION**

Freaky Fact or Fiction

113 In May 2010, a team of American paleontologists announced the discovery in Mexico of a new dinosaur species: *Coahuilaceratops magnacuerna* (kow-WHE-lah-SERA-tops mag-NA-KWER-na). This beast was a giant herbivore that lived 72 million years ago. Its most notable features are two massive horns above its eyes. At 1.2 m (4 ft) long, they are the longest horns of any known dinosaur. Scientists say little is known about the dinosaurs of Mexico, so this find is both exciting and important. What is just as exciting is that more dinosaur fossils are expected to be found in Mexico in the future.

✓ FACT   OR   FICTION

14 For 200 years, a box of ancient bones was stored in a forgotten cabinet at the Rome Dinosaur Centre. Until 2007, no-one even realised the bones were there. When they were finally rediscovered, the centre's leading scientist, Professor Mort Adello, knew it was an exciting find. DNA tests showed that this dinosaur was highly intelligent. The female of the species even allowed her young to shelter under her during storms. Professor Adello had no trouble thinking up a name for the new species. He called it *Umbrellosaurus aperiri* (um-BRELL-uh-SAWR-us AP-er-IR-ee).

 ✓ **FACT**   **OR**   **FICTION**

Freaky Fact or Fiction

115

Most teens eat a lot. They also have growth spurts when they get taller and put on weight. But there has never been a teen who could match the eating habits or the growth spurts of *Tyrannosaurus rex* (tie-RAN-uh-SAWR-us rex). Studies show that between the ages of about 14 and 18, *T. rex* gained around 2.07 kg (4.6 lb) every day! Because of its monster appetite it reached a weight of about 5 t (5.5 US t) by the time it was 20. By then its eating had slowed. It is thought to have lived to its late 20s.

✓ **FACT**   **OR**   **FICTION**

16

Imagine walking in an area so thick with dinosaur eggshells that you can't avoid stepping on one and shattering it. That's what scientists faced when they made one of paleontology's most amazing discoveries in 1998. In a desert in Argentina was a vast dinosaur nesting ground. It covered 2.6 km^2 (1 mi^2) and on it were scattered thousands of dinosaur eggs. It is thought the dinosaurs were titanosaurs that lived 70 to 90 million years ago. Fossilised skin found inside the eggs is scaly and similar to that of modern-day lizards.

 FACT **OR** **FICTION**

Freaky Fact or Fiction

117

For decades *Velociraptor* (vel-os-i-RAP-tor) has had a reputation as a particularly cruel dinosaur. It had a knife-like claw on its second toe, which many thought was used as a vicious weapon. New research shows that this claw was probably used to hold the prey rather than to kill it. An English scientist even built a mechanical copy of the claw to test how lethal it would have been. He found it could pierce skin, but would have been too blunt to tear open, or disembowel, a victim. Further studies show that the claw might have been used to help *Velociraptor* climb trees.

✓ FACT   OR   FICTION

18

elieve it or not, the mighty dinosaur was once attacked by animals about the size of squirrels. No, they weren't really brave squirrels – the dinosaurs were already dead! In 2010, American paleontologists, Nicholas Longrich and Michael J. Ryan were studying museum specimens when they found tooth marks in some dinosaur bones. They were able to tell from the teeth mark patterns that they were made by extinct squirrel-like mammals. Longrich and Ryan said it is likely the animals chewed on the bones much the same way people eat corn on the cob. The bones were more than 70 million years old – your corn cob probably won't last that long!

 FACT **OR** **FICTION**

119

A recent discovery has allowed scientists to look inside a dinosaur's brain for the first time. In 2008, a polar research team found a *Brachiosaurus* (BRACK-ee-uh-SAWR-us) buried under a mountain of ice at the South Pole. The bones of the 24-m-long (79-ft-long) beast were scattered in thousands of pieces. Its bodily organs had been eaten by polar bears. However, its head lay under a boulder so that bears could not reach it. The extreme cold protected the brain from decay. Tests are still in progress at the Albert Einstein Dinosaur Academy in Paris.

 FACT **OR** **FICTION**

Dinosaurs

Maiasaura (may-ya-SAWR-a) means 'good mother lizard'. These dinosaurs made their nests together and lived in very large herds. They dug man-sized holes into the ground to make the nests which they then filled with about 25 eggs. The *Maiasauras* grew to just over 9 m (30 ft) in length. They must have been good planners, because their nests were about this distance apart. Evidence suggests that they cared for their young, which is how they got their name.

 ✓ FACT   OR   FICTION

121

Dinosaur eggs came in all sorts of sizes and shapes. The egg of a *Velociraptor* (vel-os-i-RAP-tor) was long and thin. It had a pointed end, with lots of short ridges on the shell. It was about twice the size of a chicken egg. A *Hypselosaurus* (HIP-sih-luh-SAWR-us) egg was as big as a football and oval-shaped. The horned plant-eater, *Protoceratops* (pro-tow-SAIR-a-tops), laid a long, thin egg with bumps on it. Much larger was a sauropod (SAWR-uh-pod) egg. It was oval-shaped and the shell was covered in small bumps and short ridges.

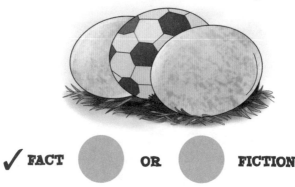

✓ **FACT**   **OR**   **FICTION**

22

ails were very important for dinosaurs. Sauropod (SAWR-uh-pod) dinosaurs, like *Diplodocus* (di-PLOD-uh-kus), often used their tails for balancing. By rearing up and leaning on their hind legs and tails, they could reach into the treetops for their food. Fast moving, two-legged dinosaurs used their tails for balance when they ran. Others used their tails for self-defence when they were under attack.

✓ **FACT**   **OR**   **FICTION**

123

All the dinosaurs that used their tails to protect themselves had four feet and ate plants. Some scientists think some dinosaurs used their tails as a kind of whip, which may have caused a stinging pain to large meat-eaters. Anklyosaurs (an-KIE-luh-SAWRS) and stegosaurs (STEG-uh-SAWRS) had clubs and spikes on their tails, which they probably used to hit their enemies and cause terrible wounds.

✓ **FACT**   **OR**   **FICTION**

'dinosaur dance floor' – that's what scientists are calling an oasis on the border of Arizona and Utah in the USA. This is where they found more than 1000 dinosaur footprints in 2008. They say the area was most likely a watering hole amid desert sand dunes during the Jurassic period, 190 million years ago. The tracks once were thought to be potholes formed by erosion. They are near a popular wind-sculpted sandstone formation known as The Wave.

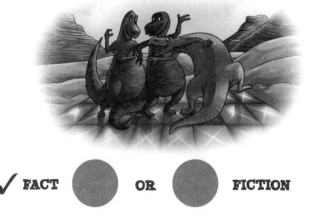

✓ FACT   OR   FICTION

125

cientists from the Albert Einstein Dinosaur Academy have released a report on their study of 'Bobo', a *Brachiosaurus* (BRACK-ee-uh-SAWR-us) found with its brain intact. A team of 50 worked on the study, and the two-year project cost US$1.9 million (A$2.2 million or £1.3 million). Using Electrical Impulse Analysis, they learnt that from 7 am to 7 pm, Bobo thought of food. The dinosaur slept for at least 12 hours, but even then his mind was active. All night long, he thought of food.

 ✓ **FACT**   **OR**   **FICTION**

26

Dinosaurs had all kinds of teeth. Some teeth were smaller, but much sharper, than human teeth. Some teeth were used to hook into the dinosaur's victims. Many teeth were huge, especially those of meat-eaters. These dinosaurs' teeth kept growing and were constantly replaced during their lives. Some of the large teeth had jagged edges, like the serrated edge of a steak knife. A large number of plant-eaters swallowed rocks, known as gizzard stones, to help them grind up plant fibres.

 FACT OR **FICTION**

Freaky Fact or Fiction

127 Fossil hunters in Brazil have discovered what could be the first complete family of dinosaurs. The bones of nine dinosaurs were found huddled together in 2009. This suggested that they may have been related. Now, through laser dating, a definite pattern has emerged. There are five young dinosaurs. These are believed to be the family's 'children'. Then there is a mature male and a mature female. These could be the parents. Finally, there is a much older male and a much older female. They are most likely the grandparents. Only one dinosaur has been named as yet. The oldest female is *Grannysaurus* (GRAN-nee-SAWR-us).

 FACT OR **FICTION**

28

Have you ever wondered just how many dinosaurs there were? So far scientists have found over 300 genera (general groupings). Within each genus there can be many different species. Most of them ate plants, and about 100 kinds ate meat. The biggest dinosaurs were plant-eaters, weighing about 90.7 t (100 US t) and measuring 33.5 m (110 ft) long. The biggest carnivores were 8 t (8.8 US t) and 13.7 m (45 ft) long.

✓ **FACT** **OR** **FICTION**

129

Triceratops (try-SAIR-a-tops), which means 'three-horned face', belonged to a group of dinosaurs known as ceratopsians (sair-a-TOP-see-ans), or horned dinosaurs. When *Triceratops* was first found in North America, it was wrongly described as a bison! *Triceratops* had three horns: one on its nose and two long ones above its eyes. It had a short neck frill and was heavily built, with strong legs. It was large and about 9 m (29.5 ft) long.

✓ **FACT**   **OR**   **FICTION**

• •

30

ome fossils are worth a lot of money, not just because they are from the time of the dinosaurs, but because they are partly made out of precious stones called opals. An opalised jawbone with teeth from *Steropodon galmani* (stair-OP-uh-don gal-MAHN-ee) was found at Lightning Ridge in New South Wales, Australia. It was one of the earliest monotreme (egg-laying mammal) fossils found in Australia. The opalised skeleton of a marine reptile, known as the 'Addyman plesiosaur' (PLEE-see-uh-SAWR), was found in 1968 by opal miners in Andamooka, South Australia.

✓ **FACT**   **OR**   **FICTION**

131

Some dinosaurs got diseases that people get today. It seems amazing, but it's true. Fossilised dinosaur bones have shown swollen areas of tumour growth. This is a sure sign that the animal suffered from cancer. In North America, a scientific team used an X-ray machine to scan 10,000 dinosaur bones. They found tumours in the bones of hadrosaurs (HAD-ruh-SAWRS). Hadrosaurs, or duck-billed dinosaurs, were plant-eaters from the Cretaceous period, about 70 million years ago.

 ✓ **FACT** **OR** **FICTION**

32

In South America today there lives a hoatzin (waht-SEEN) bird, which has claws on its wings. Baby hoatzins use their claws to climb tree branches. The prehistoric flying creature, which also had claws, was *Archaeopteryx* (ahr-kee-OP-ter-iks). *Archaeopteryx* was the earliest known bird. It appeared in the Jurassic period. It had a tooth-filled head and a long, bony tail like a dinosaur.

 ✓ **FACT**   **OR**   **FICTION**

Freaky Fact or Fiction

133 Studies of a cave deep in Egypt's Sahara Desert have shocked the scientific world. Using the latest super-density acoustic equipment – an acousticator – scientists have heard the faint but distinctive roars of a dinosaur! It is believed the beast was roaring at the very moment a tremendous sandstorm buried the cave's entrance. The dinosaur and its roar remained trapped for 150 million years. A digital music download of the roar is expected to go on sale by Christmas 2011. All proceeds will go to the Society for the Advancement of Dinosaurs.

✓ **FACT**   **OR**   **FICTION**

When scientists find fossils of dinosaur bones, they need to free the bones from the rock they are in. If they chipped them away, they might damage the bones and it would take a very long time. So the fossilised bones go to a laboratory still inside the rock. There they dip the fossil into a vat of acid. This frees the fossil from the rock without damaging it. The chemicals used are very dangerous so workers need to use protective clothing.

 ✓ FACT OR FICTION

Freaky Fact or Fiction

135

Hypsilophodon (hip-sih-LUH-foh-don) remains were first found in England in 1849. Its name means 'high crested tooth'. *Hypsilophodon* had about 30 ridged or uneven teeth which probably ground against each other, which meant they were self-sharpening. Initially some scientists thought that because of its size this plant-eating dinosaur lived in trees to protect it from bigger predators. However, it was very much a ground-dweller. At 2 m (6.5 ft) in length but only waist-height on a modern man, it was small for a dinosaur, but very quick on its feet. Because many fossils of *Hypsilophodon* have been found in one place, it is suspected that it travelled in herds.

✓ FACT   OR   FICTION

● ●

36

The Chinese have been collecting dinosaur fossils for over 2000 years. They used to refer to them as 'dragon's teeth', because it was mostly dinosaur teeth that they found. Thought to have healing powers, the bones were ground into powder. It seems likely that the mythical Chinese dragon, an important symbol in Chinese culture, might have originated from the discovery of dinosaur remains.

✓ **FACT**   **OR**   **FICTION**

Dinosaurs can be split into two groups according to their hipbones. They are the saurischians (sawr-RISS-kee-ans), or lizard-hipped dinosaurs, and the ornithischians (or-ni-THISS-kee-ans), or bird-hipped dinosaurs. All meat-eating dinosaurs were lizard-hipped, but some plant-eaters were also lizard-hipped. Ornithischians were all plant-eaters. Scientists believe that birds have evolved from lizard-hipped dinosaurs, not bird-hipped dinosaurs. Isn't that strange?

✓ FACT OR FICTION

Living in the area of the Gobi Desert in Mongolia about 80 million years ago was the *Oviraptor* (oh-vee-RAP-tor). The first skeleton of this dinosaur was found in 1923, supposedly near a nest of *Protoceratops* (pro-tow-SAIR-a-tops) eggs. So it was thought that *Oviraptor* was an egg-eater. It had an oddly shaped toothless beak and stood about 2 m (6.6 ft) tall. Its name is Greek for 'egg thief' because it was first thought that it had stolen the *Protoceratops* eggs. It is now thought the eggs probably belonged to *Oviraptor*.

 FACT **OR** **FICTION**

139

While helping to build a school vegetable garden in Queensland, Australia, a boy found remnants of a prehistoric creature. In 2010, the Year 8 boy struck what he thought was a rock, when planting vegetables. An alert teacher took a closer look and realised the rock was actually a fossilised bone from an ichthyosaur (IK-thee-uh-sawr). It was about 100 million years old. But there was still one more ancient treasure in the vegie patch. Under the bone was a book that had been borrowed from Brisbane Library in 1904!

 ✓ FACT   OR   FICTION

• •

40

nglishman Samuel Johnson wrote one of the first dictionaries. Words were Johnson's hobby, but he had another one as well. When he wasn't searching for new words, Johnson loved searching for dinosaur fossils. Every weekend for many years, he set off for the country with little more than a map, some food and a shovel. He nearly always came back empty-handed. But then, in 1723, he discovered a dinosaur that no-one had ever seen before. Johnson's son Theo was always called 'The', so he named the new dinosaur in his honour. You may have heard of a *Thesaurus* (thuh-SAWR-us).

✓ **FACT**   **OR**   **FICTION**

141

S ome people believe a dinosaur called Loch Ness lives today in a lake in Scotland. They have described the Loch Ness Monster as looking like a plesiosaur (PLEE-see-uh-sawr). This would make 'Nessy' a very old monster because plesiosaurs have been extinct for 65 million years. As well, even though plesiosaurs were water dwellers, they were air-breathing reptiles. This means that 'Nessy' would be regularly coming up for air and, no doubt, having her photo taken.

✓ **FACT**   **OR**   **FICTION**

Dinosaurs

● ● ● ● ● ● ● ● ● ● ● ● ● ● ● ● ● ● ●

Which were the fastest dinosaurs? Fossil footprints give the best information about how fast a dinosaur could move. Small dinosaurs were best suited to bursts of speeds over long distances. They usually had light skeletons and long limbs, feet, toes and claws. The small to medium coelurosaurs (SEE-loor-uh-sawrs), such as ornithomimosaurs (or-nith-uh-MIME-uh-sawrs) and dromaeosaurs (DROME-ee-uh-sawrs), were probably the fastest of all dinosaurs. Scientists have worked out that some species could run faster than 64 km/h (40 mi/h).

 FACT **OR**   **FICTION**

Freaky Fact or Fiction

143

Dinosaurs ruled the Earth for over 150 million years. This makes them the most successful group of backboned animals to have ever lived on land. Compare this with the human race – somewhere between four million and six million years! An insect that has outlived dinosaurs – and will probably outlive humans – is the cockroach.

In fact, the fossil of a giant prehistoric cockroach has been found in North America. This cockroach was scuttling around on Earth 55 million years before the dinosaurs arrived.

 ✓ FACT OR FICTION

44 arly explorers may have been the only humans to actually see living dinosaurs. In 2010, marine archeologists in Greenland found the log of an ancient Viking ship. Drawings in the log showed creatures that are almost identical to *Hadrosaurus* (HAD-ruh-SAWR-us). Professor Vivian Moss of Ivy University said this finding has excited scientists. 'This could prove that some dinosaurs survived extinction by taking to the water,' she said. 'Certainly they eventually became extinct, but they may have lived much longer than we currently think.'

 FACT **OR** **FICTION**

145

There are different kinds of fossils. Some can be animals or plants preserved in plant sap or resin (amber). Or they can be preserved in peat bogs and tar pits when they have turned into 'mummies' or have been frozen. Another type of fossil is where the living tissue has been petrified, or turned into stone. Finally, a fossil can be created when the original remains have dissolved and left a mould. This is filled with minerals such as quartz.

 ✓ **FACT** **OR** **FICTION**

● ●

46

The central region of South Africa is where many prehistoric fossils have been found. The most common plant-eaters from this area were the dicynodont (die-SIGH-no-dont). These pig-shaped mammal-like reptiles had just two canine teeth in their upper jaw. Their name means 'two dog toothed'. They ate plants with their sharp horn-lined beaks. Then they ground the food in their jaws. The dicynodont was preyed upon by the first sabre-toothed tigers, which were called gorgonopsians (GOR-gon-OP-see-ans). Scientists used to think dicynodont died out at the end of the Triassic period, but the dicynodont found in Australia may have lived alongside the dinosaurs.

✓ FACT   OR   FICTION

147

In Cretaceous times (144 million to 65 million years ago), an inland sea stretched from north to south through North America. This is called the Western Interior Seaway. Present-day fossilised dinosaur footprints show that the animals travelled along the seaway to Alaska. Some large polar dinosaurs were capable of migrating for distances up to 2600 km (1616 mi).

✓ **FACT**   **OR**   **FICTION**

48

The dinosaur with the longest neck was *Mamenchisaurus* (ma-MEN-chih-SAWR-us). Its neck was almost as long as the rest of its body, with the neck being 11 m (36 ft) of the total 25 m (82 ft) length. This one dinosaur neck was about the length of two Asian elephants in a row! *Mamenchisaurus* was an eating machine. It is thought that it stood in one place and ate everything around it, for as far as its neck would stretch – which was quite a distance. Then it would take another massive step and start eating again. *Mamenchisaurus* lived over 150 million years ago.

✓ **FACT**   **OR**   **FICTION**

149

Famous dinosaur hunter Don Duckett and his wife Daisy were on the Titanic when it sunk in 1912. The Ducketts clung to a piece of wreckage for nearly an hour and were on the point of death. They were saved when a young naval officer named Lee saw them and dragged them aboard a rescue ship. Many years later, Don Duckett repaid the favour. When he discovered a new genus of tyrannosaur (tie-RAN-uh-SAWR) in the Grand Canyon area, he named it the *Gladleesawrus* (GLAD-lee-SAWR-us.)

✓ **FACT**   **OR**   **FICTION**

50

In 1914, the residents of the small New Zealand town of Kaikoura found something very strange. The entire length of the beach was covered in fossilised dinosaur bones. Overnight there had been a massive underwater earthquake that had shaken the bones free from the ocean depths. Over many years, three dinosaurs were pieced together and are now on display in the New Zealand Museum. Kaikoura's local people collected all the small fossilised bones from the beach. As a thank-you, they were given the honour of naming the new dinosaur genus. They called it *Macropainintheneckasaurus* (MAC-row-PAIN-in-the-NECK-uh-SAWR-us).

✓ **FACT** **OR** **FICTION**

Freaky Fact or Fiction

151

The smallest dinosaur ever found was a fossilised skeleton of a *Mussaurus* (MOOSE-sawr-us). It was 20 cm (8 in) long, a little longer than a human adult's hand. *Mussaurus* lived in desert lands in South America, and its name means 'mouse lizard'. It hatched from tiny eggs 2.5 cm (1 in) long. It is estimated that it would have grown to 3 m (10 ft) long and weighed about 120 kg (260 lb).

✓ **FACT**   OR   **FICTION**

52

The Triassic period (248 to 208 million years ago) was when dinosaurs originated. It was a time of dry-climate plants and many reptiles. One of the first true flying reptiles of this time was the pterosaur (TER-uh-SAWR) called *Eudimorphodon* (YOU-die-MOR-fo-don). Found in northern Italy, it had a wingspan of about 1 m (about 3.3 ft) and lived on fish. It had a long tail with a rudder that it used for steering. Paleontologists believe that it was able to flap its wings.

✓ FACT   OR   FICTION

153

Parasaurolophus (par-a-SAWR-OL-uh-fus) had a most unusual head. Extending out from its skull was a long, tubular growth, or crest. It curved back from its snout for about 1.8 m (6 ft). It seems to have been longer in males than females and had four tubes inside it. This crest may have been used as a signal to let *Parasaurolophus* recognise one another, or to produce a sound similar to a foghorn. This animal grew to 12 m (40 ft) long and about 2.8 m (9 ft) tall at the hips. Its name means 'crested lizard'.

✓ **FACT**   **OR**   **FICTION**

● ●

54

Some medical students in the US have studied dinosaurs to help them learn about treating humans. In 2006, trainee doctors worked with paleontologists to identify a cancerous growth in a *Camptosaurus* (CAMP-tuh-SAWR-us). They used modern medical technology such as CT scans – advanced tests that use X-rays – to locate the baseball-sized tumour. The students are never likely to forget this kind of cancer again, after finding it in a dinosaur that was about 150 million years old.

The University of Pittsburgh School of Medicine and Carnegie Museum of Natural History worked in partnership on the project.

 ✓ **FACT**   **OR**   **FICTION**

Freaky Fact or Fiction

155

If you want to know what dinosaur ancestors would have looked like, a good example is *Herrerasaurus* (her-RAIR-uh-SAWR-us). It was a two-legged carnivore from Argentina, South America, which lived in the Late Triassic period, about 230 million years ago. It is estimated to have grown to about 3 to 4 m (10 to 13 ft). Its body shape suggests that it was a fast and dangerous predator, with sharp teeth and three-fingered hands. *Herrerasaurus* was named after a rancher, Victorino Herrera, who discovered it in 1958.

 ✓ FACT   OR   FICTION

56

Researchers at the Australian Dinosaurium have found two dinosaurs that were allergic to each other. Studies have shown that *Ickysaurus* (ICK-ee-SAWR-us) made *Barfasaurus* (BARF-uh-SAWR-us) physically sick. But that did not stop them liking each other. 'We have found that these two animals were inseparable,' said head researcher Dr Preston Starched. 'The constant sickness killed off *Barfasaurus*. We suspect that *Ickysaurus* simply pined away from loneliness. You could say they died for love.'

✓ **FACT**　　　**OR**　　　**FICTION**

157

ne of the most spectacular-looking dinosaurs of all was *Amargasaurus* (ah-MAHR-gah-SAWR-us). We all know that dragons never lived, but if they had, they might have looked like this creature. It had a double row of spines that ran from its neck, all along its back and halfway along its tail. It is named after La Amarga, a canyon in Argentina, where it was found in 1984. This huge herbivore grew to 10 m (33 ft) long and weighed about 5000 kg (11,000 lb).

✓ FACT   OR   FICTION

58

The biggest dinosaur-hunting trip of all took place in Africa in the early 1900s. It began when an engineer, searching for minerals in Tanzania, found pieces of gigantic fossil bones on the surface of the ground. An expedition, led by German scientists, employed over a thousand workers from 1909 to 1911. They dug up nearly 100 skeletons and hundreds of individual bones on a 3 km (1.8 mi) site. As many as 4300 loads of fossilised dinosaur bones were shipped out.

✓ FACT   OR   FICTION

Freaky Fact or Fiction

159

The scientific world was agog in late 2010 with the discovery of a startling new Australian dinosaur called *Backpackasaur* (BAK-pak-uh-sawr). The 74-million-year-old female dinosaur may be the earliest form of kangaroo. It is about the same size as a kangaroo, and it was also a hopper. Unlike the roo, the dinosaur had not one pouch, but two – one in front of its body, and one at the back. The front pouch was most likely used as a place to store food, while its young were carried in the back pouch.

✓ FACT   OR   FICTION

● ●

Can you imagine a watch made out of dinosaur poop? In 2010, Swiss watchmaker Artya and designer Yvan Arpa created such a watch. The fossilised dung used in the watch's creation came from a plant-eating dinosaur that lived in North America 100 million years ago.

The watch is self-winding and has a sapphire coating. It is water resistant and comes with a two-year warranty. Its strap is fashioned out of American cane toad skin. It is valued at US$11,290 (A$13,370 or £7715). It's not every day you can buy a watch made from fossilised dinosaur poop, right?

 FACT **OR** **FICTION**

161

Most dinosaur footprints have three toes. The other two toes were held high off the ground. Or they were lost during evolution. Tracks can provide helpful information about the behaviour of dinosaurs. The shape of a footprint shows which kind of dinosaur it was and how large it was. The space between the prints shows how fast it was moving. Multiple tracks show that a herd passed by.

 ✓ FACT OR FICTION

62

Mary Anning (1799–1847) was an early pioneer of fossil collecting. She lived all her life in Lyme Regis on the Dorset coast in England. The sea cliffs there are one of the world's most renowned places for early Jurassic marine reptile fossils. When she was only 11, Mary found her first important specimen – the whole skeleton of a large ichthyosaur (IK-thee-uh-sawr). Mary and her brother Joseph were encouraged by their father to find fossils as he then sold them to people as curiosities. Mary found the first complete plesiosaur (PLEE-see-uh-sawr) skeleton and the first British pterosaur (TER-uh-SAWR).

 FACT **OR** **FICTION**

163

The author of the novel *Jurassic Park* (from which the famous movie was made) had a dinosaur named after him. The author is Michael Crichton and the dinosaur is *Crichtonsaurus* (CRY-ton-SAWR-us). This plant-eater lived about 90 to 95 million years ago. It was a low-slung, medium-sized armoured dinosaur that lived in Asia during the Middle Cretaceous period.

✓ **FACT** OR **FICTION**

64

In 1993, *Attenborosaurus* (AT-ten-bur-row-SAWR-us) was named after the British natural history filmmaker Sir David Attenborough. *Attenborosaurus* was an aquatic reptile with a very large head and short neck. A pliosaur (PLY-uh-sawr), *Attenborosaurus* had only a few massive teeth. It is thought it used these to eat fish during the Early Jurassic period. The original fossil of *Attenborosaurus* was destroyed in a bombing raid on England during World War II. Luckily, a plaster cast had been made.

 FACT **OR** **FICTION**

165

Did you know that there is a dinosaur named after a family of famous writers? The Brontë sisters – Charlotte, Emily and Anne – all died at a young age, but not before they had written novels that eventually made them famous. In 1895, a new genus of dinosaur was found in West Yorkshire, not far from where the sisters had lived. To honour their memory, the dinosaur was called the Brontësaurus (BRON-tee-SAWR-us). This dinosaur is not related to *Brontosaurus* (BRON-toe-SAWR-us).

✓ **FACT**   OR   **FICTION**

When it was first found in the Antarctic, *Cryolophosaurus* (CRY-oh-LOAF-oh-SAWR-us) was nicknamed Elvisaurus, after the singer Elvis Presley. This was because it had a horizontal crest on top of its head that was similar to Presley's hairstyle. About 6.1 m (20 ft) long, *Cryolophosaurus* lived in the Early Jurassic period. At that time, Antarctica was not the freezing place it is today, so *Cryolophosaurus* did not need fur or feathers to keep it warm.

 FACT OR FICTION

Freaky Fact or Fiction

167 *Itchyosaur* (IT-chee-uh-sawr) is so named because it spent much of its life itching. Cellular examination of this animal's skeleton proved that its hide was infested with dandruff. Its only relief would have been to plunge into the sea and have a good long wash, but as it couldn't swim, this often proved a bad idea.

✓ **FACT** OR **FICTION**

58

olepiocephale (co-LEE-pee-oh-SEF-ah-lee) was given a funny name. *Colepio* is the Greek root for 'knuckle' and *cephale* means 'head'. Put them together and what have you got? That's right – a dinosaur named Knucklehead! A two-legged herbivore, this dinosaur was small – about 90 cm (3 ft) long. It was a type of pachycephalosaur (PACK-ee-SEF-uh-lo-sawr) or thick-headed lizard. As the name suggests, these dinosaurs had a lot of bone on top of their heads. This probably came in handy to head-butt rivals.

 FACT **OR**   **FICTION**

Freaky Fact or Fiction

169

Bambiraptor (BAM-bee-RAP-tor) sounds like it's a character from a Disney movie. But it is the real name of a dinosaur that was found by a 14-year-old boy in 1995. The boy found the near-complete skeleton of *Bambiraptor* in Montana's Glacier National Park, USA. This tiny, two-legged, bird-like raptor may have been covered with feathers. Its brain was almost as big as that of modern birds. Paleontologists study *Bambiraptor* to work out the evolutionary link between ancient dinosaurs and modern birds. And yes, it was named after the Disney Bambi character!

 ✓ **FACT** **OR** **FICTION**

Dinosaurs

●●●●●●●●●●●●●●●●●●●●●●●●●

Shuvuuia (shoe-VOO-yee-ah) is a prehistoric animal that paleontologists can't quite figure out. Was it a dinosaur or a bird? It had a small, bird-like head but dinosaur-like forelimbs with long legs and three-toed feet. It was found in Mongolia and its name is taken from the Mongolian word for 'bird'. It was about as big as a chicken and was a quick mover that probably fed on insects and worms. It's not quite as scary as some of its more famous cousins, but it shows that dinosaurs came in all shapes and sizes.

✓ **FACT**　　　**OR**　　　**FICTION**

Freaky Fact or Fiction

171 As its name suggests, *Supersaurus* (SOO-per-SAWR-us) was a massive dinosaur. It was probably as long as half a football field. However, it wasn't called *Supersaurus* just because of its size. This dinosaur could fly at great speeds, although only for short distances. As well, its armour-plated body could easily have withstood bullets. Unlike Superman, it did not have X-ray vision. In fact, it was short-sighted. But it made up for this by having quadraphonic hearing. That is, it could hear from four directions at once. This was indeed a super dinosaur.

 ✓ FACT   OR   FICTION

● ● ● ● ● ● ● ● ● ● ● ● ● ● ● ● ● ● ● ●

72

Utahraptor (YOO-tah-RAP-tor) was probably one of the best-equipped killing machines of all dinosaurs. In 1991, a well-preserved skeleton was found in Utah, USA. Utahraptor was a carnivore with large eyes, long hands and strong, clawed feet. Its main weapons were the hooked claws on its feet. While its 20-cm (8-in) claws struck out, it balanced on its tail. With this method it was a serious fighter that could kill animals much larger than itself.

✓ **FACT**   **OR**   **FICTION**

173

Researchers at the George Washington Dinosaur Centre have found evidence of a highly intelligent dinosaur. It was a rough-necked megabore (MEG-uh-BAWR) called *Bragasaurus* (BRAG-uh-SAWR-us). Deep-heat analysis of soil and bone samples proved that this reptile was a clever mimic. Head researcher Dr Allie Baba explained, '*Bragasaurus* could imitate other animals – not just frogs and birds, but maybe cockroaches, too. We still don't know why it did it, but we suspect it just liked to show off.'

✓ **FACT**   **OR**   **FICTION**

74

When he made the movie *Jurassic Park* in 1993, director Steven Spielberg perhaps did more than anyone to get people interested in dinosaurs. One of those whose fame and reputation grew because of the movie was *Velociraptor* (vel-os-i-RAP-tor). However, although they were referred to as *Velociraptor* in the movie, the dinosaurs shown were actually the much larger and more fearsome *Deinonchysus* (dye-NON-i-kus).

 ✓ **FACT**   **OR**   **FICTION**

175

One day in 1852, Ernest Fallow set off to buy an engagement ring for his girlfriend Polly. Next to the jewellery shop a very special auction was taking place. The highest bidder would win the naming rights for a new dinosaur genus. Fallow bid all the money that he had, and soon the naming rights were his. Polly was not impressed that he had named a dinosaur *Pollysaurus* (POLL-ee-SAWR-us), instead of buying a ring. It only made things worse when Ernest explained that he had done it because the dinosaur reminded him of her. Their marriage did not take place.

✓ FACT   OR   FICTION

● ● ● ● ● ● ● ● ● ● ● ● ● ● ● ● ● ●

76

The Australian airline Qantas has a dinosaur named after it – but it couldn't fly. *Qantassaurus* (KWAN-tuh-SAWR-us), which lived over 100 million years ago, was about the size of a small kangaroo but it is not thought to have been a hopping animal. Discovered in 1996, it was a herbivore that was unique to Australia. It was named after Qantas to honour the company for its work in transporting dinosaur exhibitions.

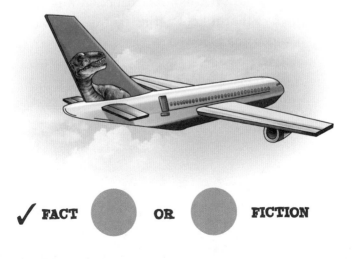

✓ **FACT** **OR** **FICTION**

177

A new device called a Brain Gauge can tell how smart dinosaurs were. Scientists say *Troödon* (TRO-uh-don) was the smartest dinosaur. It had its own primitive language. This was only a series of grunts, but it made enough sense for *Troödons* to have conversations. It also had very basic counting skills, but was probably hopeless at long division. The least smart dinosaur was *Derrosaurus* (DERR-o-SAWR-us). Its extinction was due to the fact that it kept eating its tail. It thought it was being followed.

✓ **FACT**   **OR**   **FICTION**

78

Australian scientists have discovered a dinosaur that is an ancient relative of the kangaroo. It was discovered in 2002 by an opal miner in South Australia. The dinosaur is about three times the size of a kangaroo. In all other aspects the two animals are very similar. The dinosaur even had a pouch to carry its young. It is called *Skippysaurus* (SKIPP-ee-SAWR-us).

✓ **FACT** **OR** **FICTION**

179

A bone bed discovered in Zhucheng, China in 2009 could be the last place dinosaurs gathered before their extinction. Some 15,000 bones were found stacked on top of each other at the site, which is thought to be the world's largest bone bed. Scientists say the bones date to the end of the Cretaceous period, which was the time that dinosaur extinction occurred. It is unknown yet why the dinosaurs came to this place to die.

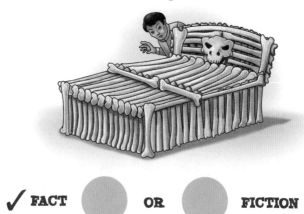

✓ **FACT**   **OR**   **FICTION**

80

In 2010, a team of international researchers made a dramatic breakthrough. While studying fossils of the oldest known 'dinobird' *Archaeopteryx* (ahr-kee-OP-ter-iks), they found chemical remains of the animal itself. Previously it was thought that the fossils were just impressions of organic material that had long ago decomposed. Powerful X-rays revealed that the fossils held fragments of actual feathers. They contained the same elements that are found in the feathers of modern-day birds. Researcher Bob Morton said, 'The discovery that certain fossils retain the chemistry of the original organisms offers scientists a new avenue for learning about long-extinct creatures.'

✓ **FACT**   **OR**   **FICTION**

Freaky Fact or Fiction

181

As a group, it is estimated that dinosaurs lived for over 160 million years. But how long did individual dinosaurs live? When the first studies on ageing were made, it was thought that dinosaurs may have lived hundreds of years. However, with new techniques come new estimates. Now scientists say that the large herbivores may have lived for about 80 years, give or take a few. Smaller, carnivorous specimens probably lived for 20 to 30 years. So it seems that eating your greens is pretty good for you.

 ✓ **FACT** **OR** **FICTION**

Dinosaurs

The fossil remains of *Explorasaurus* (ex-PLORE-uh-SAWR-us) have been found in every country of the world. It is the most travelled of all the dinosaurs. Scientists can only guess the reasons for this. Some say it was curious and liked to see what was over the next hill. Others say it was restless, forever looking for some new adventure. And there are those who believe the dinosaur simply had a poor sense of direction and was always getting lost.

 ✓ FACT **OR** **FICTION**

Freaky Fact or Fiction

183

Dinosaur eggs were once found under the New York subway. The eggs were discovered in late 1948 by a man named John Doe. He immediately rang the New York Zoo to report the find. In the meantime, his wife, Jane Doe, had cooked the eggs for breakfast, and eaten them. She immediately became ill. A doctor was quickly on the scene but, unfortunately, Mrs Doe's life was pronounced extinct.

 ✓ FACT OR FICTION

A paleontologist named Earl Douglass first discovered dinosaur fossils in Utah, USA, in 1909. He went on to find thousands more and the region he worked in became Dinosaur National Monument. Today, examples of more than half of North America's dinosaurs from the Jurassic period can be found in the park's Dinosaur National Monument Quarry. A new species of dinosaur named *Abydosaurus mcintoshi* (a-BID-oh-SAWR-us MAK-in-tosh-ee) was found in the park in 2010. *Abydosaurus* was from the sauropod (SAWR-uh-pod) family, which were huge herbivores. A dig team, who used explosives to reach the specimens, found four skulls, and two of them were intact.

✓ FACT　　OR　　FICTION

185

Epidendrosaurus (EP-ih-DEN-droh-SAWR-us) was one of the earliest 'dinobirds'. No-one knows for sure if it flew, though it's doubtful. Nevertheless, it is thought that it lived in trees and had feathers. It also had something in common with a present-day animal called the aye-aye. These rare creatures are lemurs that are only found in Madagascar. Like *Epidendrosaurus*, the aye-aye has one very long finger on each hand to allow it to dig for grubs and insects in holes and tree bark. It's very likely that this is what *Epidendrosaurus* did as well.

✓ **FACT**   **OR**   **FICTION**

36

Carnotaurus (KAR-no-TAWR-us) was an odd-looking dinosaur from South America. It had a small skull for such a big animal, and it also had tiny arms. Its name means 'meat-eating bull', so if you guessed that it was a carnivore that looked a bit like a bull, you were right. It is best known because it had bull-like horns above each eye. It was around 7.6 m (25 ft) long and weighed about 1 t (1.1 US ton).

 ✓ **FACT**   **OR**   **FICTION**

Freaky Fact or Fiction

187

The first dinosaur skeleton ever to be put together was truly a jigsaw. For nearly 200 years the credit for assembling the dinosaur has been given to famed dinosaur hunter Lord Claude Broad. Now, with the release of his lost journal, the truth has finally been revealed. Lord Broad invited the members of the Surrey Jigsaw Club over for tea. Afterwards, he challenged them to the ultimate jigsaw puzzle. The members took just two days to connect more than 40,000 pieces of fossilised dinosaur bones.

 FACT OR **FICTION**

● ●

38

If you look at the skeleton of a *Tuojiangosaurus* (too-HWANG-oh-SAWR-us) you might just start to believe in dragons. The partial remains of two of these dragon look-alikes have been found in Sichuan Province, China. Records show that many hundreds of years ago villagers collected fossils from here and sold them as dragon bones. These were highly prized for use in traditional medicine. *Tuojiangosaurus* was a prickly customer. It had large spikes over most of its body, including its tail and shoulders. It was a herbivore that was about 7 m (23 ft) long. Its name means 'Tuo River lizard'.

✓ **FACT**   **OR**   **FICTION**

Freaky Fact or Fiction •••••••••••

189

It is clear that dinosaurs migrated in vast numbers. Across western North America massive 'bone beds' of several species have been found. These dinosaur graveyards can hold the remains of hundreds or thousands of beasts. It makes sense that herds of this size couldn't stay in the same place for too long, because they would soon exhaust food supplies. They had to keep moving.

 ✓ FACT   OR   FICTION

Dinosaurs

In 2010, a yoga instructor in Argentina found herself in an amazing position. While performing her exercises she accidentally came across some very big footprints. Scientists found that the tracks, which were in good condition and up to 1.2 m (4 ft) in diameter, had been made more than 90 million years ago by sauropod (SAWR-uh-pod) dinosaurs. The find was in an area that is known as Argentina's Jurassic Park. In 1993, the remains of a *Giganotosaurus* (JI-ga-NO-to-SAWR-us), the largest carnivorous dinosaur in the world, were found there.

 ✓ **FACT**   **OR**   **FICTION**

Freaky Fact or Fiction

191 In 2009, US scientists discovered one of the world's largest dinosaur 'graveyards'. In a paper released by the Boston Academy of Advanced Dinosauria (BAAD), it is revealed that close to one million dinosaurs died at the site in Nevada. One of the most interesting features of the site is that large boulders found nearby were actually made of coprolite, or petrified dinosaur dung. Incredibly, when the outer layer of the boulders was chipped away, the smell of the dung was still strong.

 FACT **OR**   **FICTION**

92

In 1995, while studying meteorites that had fallen in Arizona, a research team found minute evidence of dinosaur fossils. Professor Miriam Crock, head researcher at the New York Meteorium, said that it is 'highly unlikely' that the find could mean dinosaurs once existed on other planets. 'There is no doubt that the particles we detected came from a *stegosaur* (STEG-uh-SAWR),' she said. 'However, as scientists, we believe there has to be a logical reason for this. It could be that some freak atmospheric condition caused the fossil tissue to be embedded in a meteorite.'

✓ FACT   OR   FICTION

193

Sometimes just one surviving fossil is all there is to show for a whole species of dinosaur. This is the case with *Eustreptospondylus oxoniensis* (yoo-strep-toe-SPON-die-lus OX-on-ee-EN-sis). The one and only specimen was found in a quarry in Oxford, England, and was named by Richard Owen, the man who first coined the name 'dinosaur'. The specimen was originally about 5 m (16.4 ft) long and weighed 500 kg (1102 lb); however, it was probably a young dinosaur that was still growing.

 FACT **OR** **FICTION**

It seems that dinosaurs may have been the first lovers of fast food. With their very long necks, herbivores were able to stand in one place and simply swivel their necks about to find what was on the plant menu. Researchers from the University of Bonn in Germany, led by Professor Martin Sander, have found that dinosaurs gulped their food instead of chewing it. Chewing takes time.

An animal as large as a dinosaur needed to eat quickly in order to maintain its energy demands. Professor Sander said this explained why dinosaurs grew to be so large.

 ✓ **FACT**   **OR**   **FICTION**

Freaky Fact or Fiction ● ● ● ● ● ● ● ● ● ● ●

195

A dog named Blinky once 'killed' a *Tyrannosaurus rex* (tie-RAN-uh-SAWR-us rex). In 1986, Blinky broke away from its owner while out walking in London's Hyde Park. The dog kept running until it reached the Queen Victoria Dinosaur Museum. Slipping through a back door, he found himself surrounded by towering dinosaurs. Blinky loved bones, so he just had to have one. Before security guards could stop him, he dived on a *T. rex* and pulled a thigh bone free. The *T. rex* collapsed into a million broken pieces. Blinky is now in the record books as the only dog to destroy a dinosaur.

✓ **FACT**   **OR**   **FICTION**

96

orosaurus (TOR-uh-SAWR-us) was discovered in North America in 1891. It is not named after a bull, as the prefix *toro* suggests. Its name means 'pierced lizard'. It was called this because it has large holes in its skull. A herbivore, it had a very large frill, or skull plate, and two fearsome horns. *Torosaurus* probably reached a length of 7.5 m (25 ft) and weighed about 3.6 t (4 US t). It was long considered to be related to *Triceratops* (try-SAIR-a-tops) but in 2010 the discovery was made that *Triceratops* is just a juvenile *Torosaurus*! So, the *Torosaurus* became extinct for a second time when it was reclassified.

 FACT **OR** **FICTION**

Freaky Fact or Fiction

197

In 1972, a science teacher named Al Lakusta went on a picnic at Pipestone Creek, in Alberta, Canada. He noticed something that looked like brown fossilised rib fragments. Mr Lakusta had stumbled on one of the world's richest dinosaur bone beds. As large as a football field and more than 70 million years old, it was the last resting place of a rare dinosaur, which was partly named in the teacher's honour. It is called *Pachyrhinosaurus lakustai* (PAK-ee-rye-no-SAWR-us la-KUS-tye). By 2010, more than a dozen skulls had been taken from the site. There is still much more excavation to be done.

✓ **FACT**   **OR**   **FICTION**

Scientists have put together a theory about the origin of massive dinosaur bone beds, such as the one in Pipestone Creek, Canada. They say it is likely that herds of migrating dinosaurs were crossing a flooding river, when many of them were swept away and killed. Scavenger dinosaurs picked up the smell of the rotting carcasses and went into the river, only to die themselves. This pattern continued, creating huge bone beds.

✓ FACT   OR   FICTION

199

ew studies show that dinosaur soup could hold the secret to weight loss. The bones are crushed and deep-frozen before being made into a soup. British woman Twiggy Reed says she lost half her body weight after going on the dinosaur diet. 'When you eat something that's millions of years old you usually expect it will be kind of stale,' she said. 'But it wasn't. Just add a little salt and pepper and it's lovely.'

✓ **FACT**   OR   **FICTION**

Tyler Lyson was 16 when he found his mummy. No, not his mother, his dinosaur mummy! In 1999, Lyson found a huge herbivore called *Edmontosaurus* (ed-MON-tuh-SAWR-us). Not only bones, but also fossilised skin, ligaments and tendons were identified. Nicknamed Dakota, because it was found in the US state of North Dakota, the dinosaur is said to be one of the best-preserved dinosaurs ever found. In fact, one leading scientist has stated it may be the closest we'll ever come to seeing a living dinosaur.

✓ **FACT**   **OR**   **FICTION**

Freaky Fact or Fiction

201

Scientists are confident that one day dinosaurs will live again. Professor Matt Hari of Japan's Dinosaur Institute said the technology was already in place for bringing back the ancient reptile. 'All we need is more DNA,' he said. 'At present our vats are about a third full. Each time a dinosaur is found we get a little more. My guess is that in 10 to 20 years the dinosaur will be back.'

 ✓ **FACT**   **OR**   **FICTION**

● ●

When a new discovery of a dinosaur is made it is very much like finding a needle in a haystack. After all, they have been extinct for 65 million years. Over 700 species of dinosaur have been named so far. It is estimated that at least this many again are still out there, waiting to be found. But this is a very small figure when compared to other creatures. For instance, there are about 10,000 known bird species, 20,000 fish species and as for insects, well, there are a lot. Estimates range from 1.5 million to 30 million.

✓ **FACT**   **OR**   **FICTION**

203

Psittacosaurus (SIT-uh-ko-SAWR-us) was on the small side for a dinosaur. When it stood up on its back legs it was still only about 1.2 m (4 ft) tall. It was roughly 2 m (6.5 ft) long. Its odd name means 'parrot lizard'. That's because its head resembled a parrot's. A herbivore and probably a fast mover, it has been found in China, Mongolia and Thailand. Scientists think that, like the parrot, this dinosaur ate seeds and nuts, and most likely cracked the nuts with its tough beak.

✓ **FACT** **OR** **FICTION**

04

The dinosaur called *Styrocosaurus* (sty-ROW-kuh-SAWR-us) could best be described as a very odd mix. It had several long horns jutting out above its frill. On its snout was another horn, very similar to that of the present-day rhinoceros. And, like *Psittacosaurus* (SIT-uh-ko-SAWR-us), its snout was shaped like a parrot's beak. A herbivore, it was about as tall as a rhino and grew to about 5.2 m (17 ft) long. It could weigh as much as 2.7 t (3 US t).

 FACT **OR** **FICTION**

Answers

1. Fact.

2. Fact.

3. Fact.

4. Fact.

5. Fact.

6. Fact.

7. **Fiction.**

8. Fact.

9. Fact.

10. Fact.

11. **Fiction.**

12. Fact.

13. **Fiction**, but you may have heard of a dog breed called the golden retriever.

14. **Fiction.**

15. Fact.

16. Fact.

17. Fact.

18. **Fiction.**

19. Fact.

20. Fact.

21. Fact.

22. Fact.

23. **Fiction.**

24. Fact.

25. Fact.

26. Fact.

27. Fact.

28. **Fiction.**

29. Fact.

30. Fact.

31. **Fiction.**

32. Fact.

33. Fact.

34. **Fiction.**

35. Fact.

36. Fact.

37. **Fiction.**

38. Fact.

39. Fact.

40. Fact.

41. Fact.

42. Fiction.

43. Fact.

44. Fact.

45. Fact.

46. Fiction, but Elvis Presley was born in Tupelo, Mississippi.

47. Fact.

48. Fact.

49. Fact.

50. Fact.

51. Fiction. *Dollodon* was really a dinosaur, but it was not made into a toy, nor did the Transylvanians name it.

52. Fact.

53. Fact.

54. Fact.

55. Fact.

56. Fiction. *Afrovenator* was a dinosaur, but was not named after an African inventor.

57. Fact.

58. Fact.

59. Fact.

60. Fiction. All information here is fictional. However, there was a dinosaur with a similar name, the *Chungkingosaurus*.

61. Fact.

62. Fact.

63. Fact.

64. Fact.

65. Fact.

66. Fact.

67. Fact.

68. Fiction, but there is a dinosaur called Bambiraptor.

69. Fact.

Answers

70. Fact.

71. Fact.

72. Fact.

73. Fact.

74. Fiction.

75. Fact.

76. Fact.

77. Fiction. *El Dorado* is the name of a legendary lost city of gold. Dinosaurs are not part of the legend. However, there were gold rushes in the Klondike and Yukon, and *El Dorado* does mean 'the golden one' in Spanish.

78. Fact.

79. Fact.

80. Fact.

81. Fact.

82. Fiction.

83. Fiction.

84. Fiction.

85. Fact.

86. Fact.

87. Fact.

88. Fact.

89. Fiction.

90. Fact.

91. Fact.

92. Fact.

93. Fact.

94. Fiction, although there is evidence that dinosaurs were migratory animals.

95. Fiction. Please note that Horace Rubble is no relation to Barney Rubble, best friend of Fred Flintstone, from the movie and television show *The Flintstones*.

96. Fact.

97. Fact.

98. Fact.

99. Fact.

100. Fiction.

101. Fact.

102. Fact.

103. Fact.

104. Fiction.

105. Fact.

106. Fact.

107. Fact.

108. Fiction.

109. Fact.

110. Fact.

111. Fact.

112. Fact.

113. Fact.

114. Fiction.

115. Fact.

Answers

116. Fact.

117. Fact.

118. Fact.

119. **Fiction**, and there are no polar bears at the South Pole.

120. Fact.

121. Fact.

122. Fact.

123. Fact.

124. Fact.

125. **Fiction.**

126. Fact.

127. **Fiction.**

128. Fact.

129. Fact.

130. Fact.

131. Fact.

132. Fact.

133. **Fiction.**

134. Fact.

135. Fact.

136. Fact.

137. Fact.

138. Fact.

139. **Fiction.** Though a Queensland schoolboy did find ichthyosaur remains while working in a school vegetable patch in 2010, there was no library book.

140. **Fiction**, but Samuel Johnson did write one of the first dictionaries.

141. Fact.

142. Fact.

143. Fact.

144. **Fiction.**

145. Fact.

146. Fact.

147. Fact.

148. Fact.

149. **Fiction.**

150. **Fiction.**

151. Fact.

152. Fact.

153. Fact.

154. Fact.

155. Fact.

156. Fiction.

157. Fact.

158. Fact.

159. Fiction.

160. Fact.

161. Fact.

162. Fact.

163. Fact.

164. Fact.

165. Fiction, but the Brontës were famous writers.

Answers

166. Fact.

167. Fiction. However, ichthyosaurs are marine reptiles from the time of the dinosaurs that have a similar name.

168. Fact.

169. Fact.

170. Fact.

171. Fiction. Though there is a dinosaur called *Supersaurus,* it didn't have super powers.

172. Fact.

173. Fiction.

174. Fact.

175. Fiction.

176. Fact.

177. Fiction, but *Troödon* is considered to have been the smartest dinosaur.

178. Fiction.

179. Fact.

180. Fact.

181. Fact.

182. Fiction.

183. Fiction.

184. Fact.

185. Fact.

186. Fact.

187. Fiction.

188. Fact.

189. Fact.

190. Fact.

191. Fiction. However, some fossilised dung does still smell after thousands of years. One such case is the giant ground sloth from North America; it was dead 15,000 years, but was still smelly!

192. Fiction.

193. Fact.

194. Fact.

195. **Fiction.**

196. Fact.

197. Fact.

198. **Fiction.**

199. **Fiction.**

200. Fact.

201. **Fiction.**

202. Fact.

203. Fact.

204. Fact.

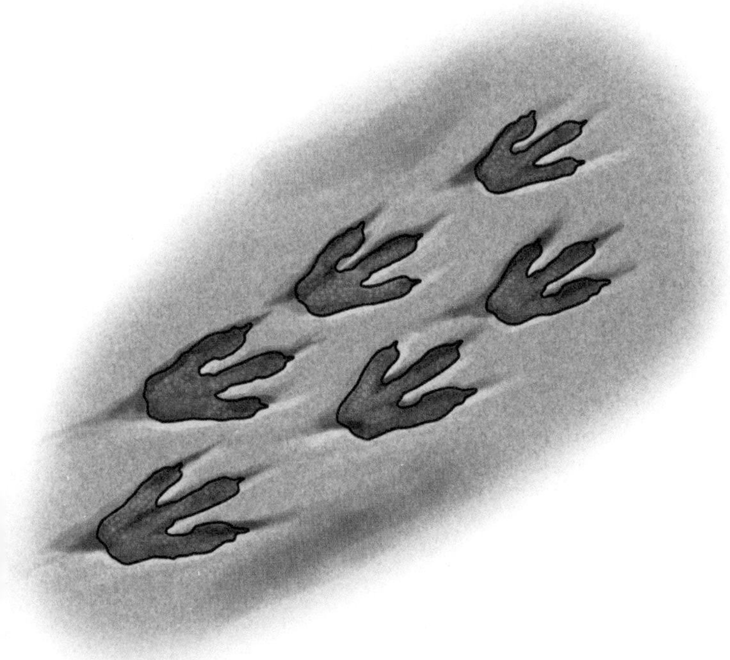

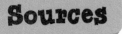

Sources

1. Michael Benton, *The Penguin Historical Atlas of the Dinosaurs*, 1996

2. Michael Benton, *The Penguin Historical Atlas of the Dinosaurs*, 1996; US Geological Survey, http://pubs.usgs.gov, 2001; Encyclopaedia Britannica Online, www.britannica.com, 2010

3. Michael Benton, *The Penguin Historical Atlas of the Dinosaurs*, 1996; US Geological Survey, http://pubs.usgs.gov, 2001; Encyclopaedia Britannica Online, www.britannica.com, 2010

4. Michael Benton, *Kingfisher Factfinder: Dinosaurs* (book), 2001; Natural History Museum, www.nhm.ac.uk, 2010; Encyclopaedia Britannica Online, www.britannica.com, 2010

5. *Guinness World Records 2009* (book), 2009

6. *Guinness World Records 2009* (book), 2009

7. Fiction.

8. *Scholastic Australian & World Records 2009* (book), 2009

9. 'Mystery Solved! Giant Asteroid Killed the Dinosaurs, say Scientists', NYDailyNews.com, www.nydailynews.com, 2010

10. Encyclopædia Britannica Online, www.britannica.com, 2010

11. Fiction.

12. Enchanted Learning, www.enchantedlearning.com, 2010

13. Fiction.

14. Fiction.

15. Michael Benton, *Kingfisher Factfinder: Dinosaurs* (book), 2001; Answers.com: ReferenceAnswers, www.answers.com, 2010; Enchanted Learning, www.enchantedlearning.com, 2010

16. Stanford University News Services, http://news.stanford.edu, 1996; Enchanted Learning, www.enchantedlearning.com, 2010

17. '10 Facts About Tyrannosaurus Rex', Bob Strauss, About.com, http://dinosaurs.about.com, 2010

18. Fiction.

19. Brian Cooley and Mary Ann Wilson, *Make-A-Saurus: My Life with Raptors and Other Dinosaurs* (book), 2000; David Norman and Angela Milner, *Collins Eyewitness Guides: Dinosaurs* (book), 1991; Benjamin Waterhouse Hawkins, www.bwaterhousehawkins.com

20. AgeofDinosaurs.com, http://ageofdinosaurs.com; Scott Hocknull and Dr Alex Cook, *Amazing Facts about Australian Dinosaurs* (book), 2006

21. Michael Benton, *Kingfisher Factfinder: Dinosaurs* (book), 2001; Oceans of Kansas Paleontology, www.oceansofkansas.com, 2009

22. Michael Benton, *Kingfisher Factfinder: Dinosaurs* (book), 2001; Michael Benton, *The Penguin Historical Atlas of the Dinosaurs*, 1996; Gondwana Studios, www.gondwanastudios.com, 2010

23. Fiction.

24. Australian Museum, http://australianmuseum.net.au, 2009; Enchanted Learning, www.enchantedlearning.com, 2010; Dinosaurs for Kids, www.kidsdinos.com, 2010; AgeofDinosaurs.com, http://ageofdinosaurs.com

25. Enchanted Learning, www.enchantedlearning.com, 2010; Dinosaurs for Kids, www.kidsdinos.com

26. Michael Benton, *Kingfisher Factfinder: Dinosaurs* (book), 2001; Michael Benton, *The Penguin Historical Atlas of the Dinosaurs*, 1996; Strange Science, www.strangescience.net.htm, 2009

27. Michael Benton, *Kingfisher Factfinder: Dinosaurs* (book), 2001; Encyclopaedia Britannica Online, www.britannica.com, 2010

28. Fiction.

29. ABC News, Australia, www.abc.net.au, 2007; Cosmos, www.cosmosmagazine.com, 2007; The Hairy Museum of Natural History, www.hmnh.org, 2008

30. Enchanted Learning, www.enchantedlearning.com, 2010; HowStuffWorks.com, http://animals.howstuffworks.com, 2008; Michael Benton, *Kingfisher Factfinder: Dinosaurs* (book), 2001

31. Fiction.

32. Enchanted Learning, www.enchantedlearning.com, 2010; Dr David Norman, *The Illustrated Encyclopedia of Dinosaurs* (book), 1985; Science Museum of Minnesota: Science Buzz, www.sciencebuzz.org, 2008

33. Michael Benton, *Kingfisher Factfinder: Dinosaurs* (book), 2001; 'Alamosaurus', Bob Strauss, About.com, http://dinosaurs.about.com, 2010; National Park Service, US Department of the Interior, www.nps.gov, 2010

34. Fiction.

35. Michael Benton, *Kingfisher Factfinder: Dinosaurs* (book), 2001; Walking with Dinosaurs (ABC, BBC), http://abc.net.au, 1999; New Mexico Museum of Natural History and Science, http://nmstatefossil.org, 2009

36. HowStuffWorks.com, http://animals.howstuffworks.com, 2010; Dinosaurs for Kids, www.kidsdinos.com, 2010; AgeofDinosaurs.com, http://ageofdinosaurs.com; 'Saichania', Bob Strauss, About.com, http://dinosaurs.about.com, 2010

37. Fiction.

38. Walking with Dinosaurs (ABC, BBC), www.abc.net.au, 1999; Michael Benton, *Kingfisher Factfinder: Dinosaurs* (book), 2001; 'Quetzalcoatlus', Bob Strauss, About.com, http://dinosaurs.about.com, 2010

39. 'Untouched on a Shelf for 113 Years: A Dusty Bone of the Dinosaur No One Knew Existed', Guardian News and Media, www.guardian.co.uk, 2007; Mike Taylor, www.miketaylor.org.uk, 2007

40. Alaska Museum of Natural History, www.alaskamuseum.org, 2005; Encyclopaedia Britannica Online, www.britannica.com, 2010

41. 'The 10 Smartest Dinosaurs', About.com, http://dinosaurs.about.com, 2010

42. Fiction.

43. US Geological Survey, http://pubs.usgs.gov.html, 2001; Geology Shop, www.geologyshop.co.uk, 2002; Life's Little Mysteries, www.lifeslittlemysteries.com, 2010

44. Walking with Dinosaurs (ABC, BBC), www.abc.net.au, 1999; Michael Benton, *Kingfisher Factfinder: Dinosaurs* (book), 2001

45. Australian Age of Dinosaurs, www.aaodl.com, 2009; Queensland Museum, www.qm.qld.gov.au, 2010

46. Fiction.

47. David Norman and Angela Milner, *Collins Eyewitness Guides: Dinosaurs* (book), 1991; Enchanted Learning, www.enchantedlearning.com, 2010

48. Dinosaur Facts, www.dinosaurfact.net, 2010; Paleos: The History of Life on Earth, www.palaeos.com, 2005; National History Museum, www.nhm.ac.uk, 2007

49. Michael Benton, *Kingfisher Factfinder: Dinosaurs* (book), 2001; Enchanted Learning, www.enchantedlearning.com, 2010

50. Michael Benton, *Kingfisher Factfinder: Dinosaurs* (book), 2001; Answers.com, www.answers.com, 2010; 'Dimorphodon', Bob Strauss, About.com, http://dinosaurs.about.com, 2010

51. Fiction.

52. Enchanted Learning, www.enchantedlearning.com, 2010; 'How Dinosaurs Are Named', Bob Strauss, About.com, http://dinosaurs.about.com, 2010

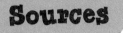
Sources

53. Dinosaur Facts, www.dinosaurfact.net, 2010; 'Chirostenotes', Bob Strauss, About.com, http://dinosaurs.about.com, 2010

54. Enchanted Learning, www.enchantedlearning.com, 2010

55. Museum Victoria, http://museumvictoria.com.au; My Jurassic Park, www.myjurassicpark.com, 2006

56. Fiction.

57. 'True-Color Dinosaur Revealed: First Full-Body Rendering', National Geographic, http://news.nationalgeographic.com, 2010

58. Hooper Virtual Natural History Museum, Ottawa–Carleton Geoscience Centre, http://hoopermuseum.earthsci.carleton.ca, 1998; Enchanted Learning, www.enchantedlearning.com, 2010; 'Dsungaripterus', Bob Strauss, About.com, http://dinosaurs.about.com, 2010

59. Michael Benton, *Kingfisher Factfinder: Dinosaurs* (book), 2001; Benedictine University, www.ben.edu, 2010; HowStuffWorks.com, http://animals.howstuffworks.com, 2010

60. Fiction.

61. Dinosaur Den, www.dinosaurden.co.uk, 2003; 'Dinosaur of the Day – Gargoyleosaurus', Bob Strauss, About.com, http://dinosaurs.about.com, 2009; Enchanted Learning, www.enchantedlearning.com, 2010

62. Encyclopædia Britannica Online, www.britannica.com, 2010; HowStuffWorks, http://animals.howstuffworks.com, 2010; Canadian Museum of Nature, http://nature.ca, 2010

63. Dinosaurs for Kids, www.kidsdinos.com, 2010; GEOERA: The World of Dinosaurs, www.dinosaurusi.com; Encyclo: Online Encyclopedia, www.encyclo.co.uk, 2010; AgeofDinosaurs.com, http://ageofdinosaurs.com.htm

64. National History Museum, www.nhm.ac.uk, 2007; 'Telmatosaurus', Bob Strauss, About.com, http://dinosaurs.about.com, 2010; DinoDictionary.com, www.dinodictionary.com, 2005

65. Museum Victoria, http://museumvictoria.com.au

66. Michael Benton, *Kingfisher Factfinder: Dinosaurs* (book), 2001; University of California Museum of Paleontology, www.ucmp.berkeley.edu, 2005

67. 'Digging Up Fossils', Scholastic: Teachers, www2.scholastic.com, 1988

68. Fiction.

69. Enchanted Learning, www.enchantedlearning.com, 2010; 'Zuniceratops', Bob Strauss, About.com, http://dinosaurs.about.com.htm, 2010

70. Children's Museum of Indianapolis, www.childrensmuseum.org, 2010; British Council, LearnEnglish Central, www.britishcouncil.org.htm

71. Australian Museum, http://australianmuseum.net.au, 2009; Museum of Nature & Science Dallas, Texas, www.natureandscience.org, 2010

72. 'Dinosaur Had Crocodile-Like Skull', National Geographic News, http://news.national-geographic.com.html, 2008; Michael Benton, *Kingfisher Factfinder: Dinosaurs* (book), 2001; 'Dinosaur directory: Baryonyx', guardian.co.uk, www.guardian.co.uk, 2006

73. 'The First Mammals – Mammals of the Triassic, Jurassic and Cretaceous Periods', Bob Strauss, About.com, http://dinosaurs.about.com, 2010; Weber State University, Utah, http://faculty.weber.edu, 2010

74. Fiction.

75. American Museum of Natural History, www.amnh.org, 2006

76. Michael Benton, *Kingfisher Factfinder: Dinosaurs* (book), 2001; Answers.com: ReferenceAnswers, www.answers.com, 2010; The TalkOrigins Archive, www.talkorigins.org, 2006

77. Fiction.

78. Michael Benton, *Kingfisher Factfinder: Dinosaurs* (book), 2001

79. 'Fossil of Dog-Sized Horned Dinosaur Unearthed in China', National Geographic News, http://news.nationalgeographic.com, 2002; 'New Dinosaur Related to *Triceratops*', Science Daily, www.sciencedaily.com, 2002; California Academy of Sciences, www.calacademy.org, 2005

80. '*Tyrannosaurus rex* Was a Slowpoke', National Geographic News, http://news.nationalgeographic.com, 2002

81. A History of Dinosaur Hunting and Reconstruction, www.dinohunters.com, 2007; University of California Museum of Paleontology, www.ucmp.berkeley.edu, 2010

82. Fiction.

83. Fiction.

84. Fiction.

85. 'Revealed: Australia's Very Own Little Tyrannosaur', *The Sydney Morning Herald* (newspaper), Sydney, Australia, 26 March 2010

86. 'Snake Preyed on Baby Dinosaurs', *The Sydney Morning Herald* (newspaper), Sydney, Australia, 3 March 2010

87. 'Nothing Dull about an Orange Dinosaur', *The Sydney Morning Herald* (newspaper), Sydney, Australia, 28 January 2010

88. 'Dinosaurs "Lucky" Crurotarsan Predators Killed off by Volcanoes, Study Finds', News.com.au, www.news.com.au, 2010

89. Fiction.

90. Encyclopædia Britannica Online, www.britannica.com, 2010

91. Encyclopædia Britannica Online, www.britannica.com, 2010

92. Enchanted Learning, www.enchantedlearning.com, 2010

93. Enchanted Learning, www.enchantedlearning.com, 2010

94. Fiction.

95. Fiction.

96. 'Iguanodon', Beverly Eschberger, Suite101.com, www.suite101.com, 2000; Enchanted Learning, www.enchantedlearning.com, 2010

97. 'Runway Found for Flying Reptiles', BBC News, http://news.bbc.co.uk, 2009

98. *Encarta Encyclopedia* (CD), 1999

99. 'Titanosaurs – The Last of the Sauropods', Bob Strauss, About.com, http://dinosaurs.about.com, 2010; Queensland Museum, www.qm.qld.gov.au, 2010; 'Found! Australia's Largest Dinosaurs', ABC News in Science, www.abc.net.au, 2007

100. Fiction.

101. Steve Parker, *The Age of Dinosaurs: Dinosaurs and Birds,* (book), 2000; ABC Science, www.abc.net.au, 1998

102. Michael Benton, *Kingfisher Factfinder: Dinosaurs* (book), 2001; Dinosaurs for Kids, www.kidsdinos.com, 2010; Steve Parker, *The Age of Dinosaurs: Dinosaurs and Birds,* (book), 2000

103. Steve Parker, *The Age of Dinosaurs: Dinosaurs and Birds*, (book), 2000; Enchanted Learning, www.enchantedlearning.com, 2010; Michael Benton, *Kingfisher Factfinder: Dinosaurs* (book), 2001

104. Fiction.

105. Adam Hibbert, *Dangerous Dinosaurs* (book), 2006; The Dinosaur Interplanetary Gazette, www.dinosaur.org, 2006

106. Mick Manning and Brita Granström, *Dinomania: Things to Do with Dinosaurs* (book), 2001; 'The Poop on Dinos – Fossilized Dinosaur Dung – Includes Information on Coprolite', *Science World* (magazine), http://findarticles.com, United States of America, October 1998

107. *First Dinosaur Encyclopedia*, 1997; *The Penguin Historical Atlas of the Dinosaurs*, 1996

108. Fiction.

109. *First Dinosaur Encyclopedia*, 1997; Scott Hocknull and Dr Alex Cook, *Amazing Facts about Australian Dinosaurs* (book), 2006; Museum Victoria, http://museumvictoria.com.au, 2010

110. Adam Hibbert, *Dangerous Dinosaurs* (book), 2006; 'Pterodaustro', Bob Strauss, About.com, http://dinosaurs.about.com, 2010

111. 'Oldest Dinosaur Burrow Discovered', BBC Earth News, http://news.bbc.co.uk, 2009

112. 'Feathers Tied to Origin of Dinosaurs', MSNBC Digital Network: Science and Technology, www.msnbc.msn.com, 2009

113. 'First Horned Dinosaur from Mexico: Plant-Eater Had Largest Horns of Any Dinosaur', Science Daily, www.sciencedaily.com, 2010

114. Fiction.

115. 'Teenage *T. Rex's* Monstrous Growth', BBC One-Minute World News, http://news.bbc.co.uk, 2004

116. 'Dinosaur "Lost World" Discovered', BBC News Online, http://news.bbc.co.uk, 1998; 'First Dinosaur Embryo Skin Discovered', Science A Go Go, www.scienceagogo.com, 1998

117. 'Dino Reputation "Is Exaggerated"', BBC News, http://news.bbc.co.uk, 2005; 'Velociraptor's "Killing" Claws Were for Climbing', New Scientist, www.newscientist.com, 2009

118. 'Dinosaur-Chewing Mammals Leave Behind Oldest Known Tooth Marks', Science Daily, www.sciencedaily.com, 2010

119. Fiction.

120. Enchanted Learning, www.enchantedlearning.com, 2010

121. John Long, *Insiders: Dinosaurs* (book), 2007; David Norman and Angela Milner, *Collins Eyewitness Guides: Dinosaurs* (book), 1991

122. David Norman and Angela Milner, *Collins Eyewitness Guides: Dinosaurs* (book), 1991; Planet Dinosaur, http://planetdinosaur.com, 2010

123. David Norman and Angela Milner, *Collins Eyewitness Guides: Dinosaurs* (book), 1991; Planet Dinosaur, http://planetdinosaur.com, 2010

124. 'A Dinosaur Dance Floor', EurekAlert, www.eurekalert.org, 2008; 'Dinosaur "Dance Floor" Found in Arizona', National Geographic News, http://news.nationalgeographic.com, 2008

125. Fiction.

126. David Norman and Angela Milner, *Collins Eyewitness Guides: Dinosaurs* (book), 1991

127. Fiction.

128. 'Dinosaur Diet', Scholastic: Teachers, http://content.scholastic.com, 2010

129. David Norman and Angela Milner, *Collins Eyewitness Guides: Dinosaurs* (book), 1991; Michael Benton, *Kingfisher Factfinder: Dinosaurs* (book), 2001

130. Scott Hocknull and Dr Alex Cook, *Amazing Facts about Australian Dinosaurs* (book), 2006; Australian Museum, http://australianmuseum.net.au, 2009

131. Michael Benton, *Kingfisher Factfinder: Dinosaurs* (book), 2001; David Norman and Angela Milner, *Collins Eyewitness Guides: Dinosaurs* (book), 1991; 'Dinosaurs Got Cancer', BioEd Online, www.bioedonline.org, 2003

132. David Norman and Angela Milner, *Collins Eyewitness Guides: Dinosaurs* (book), 1991

133. Fiction.

134. David Norman and Angela Milner, *Collins Eyewitness Guides: Dinosaurs* (book), 1991; Helen Fields: Science Writer, http://heyhelen.com, 2010

135. Dinosaur Den, www.dinosaurden.co.uk, 2003

136. David Norman and Angela Milner, *Collins Eyewitness Guides: Dinosaurs* (book), 1991; 'Are Chinese Dinosaurs Dragons or Fakes?, Associated Content, www.associatedcontent.com, 2008; 'Dragons, Dinosaurs, and "Fiery Serpents"', Apologetics Press, www.apologeticspress.org, 2003

137. Caroline Bingham, *First Dinosaur Encyclopedia*, 2007; National Park Service, US Department of the Interior, www.nps.gov, 2004

138. Caroline Bingham, *First Dinosaur Encyclopedia*, 2007; Michael Benton, *Kingfisher Factfinder: Dinosaurs* (book), 2001

139. 'Ichthyosaur Found in School Vegie Patch', ABC News, www.abc.net.au, 2010

140. Fiction.

141. Legend of Nessie Official Website, www.nessie.co.uk, 2010; The Plesiosaur Site, www.plesiosaur.com.php, 2009

142. Scott Hocknull and Dr Alex Cook, *Amazing Facts about Australian Dinosaurs* (book), 2006; Michael Benton, *Kingfisher Factfinder: Dinosaurs* (book), 2001

143. John Long, *Dinosaurs of Australia* (book) 1989; Caroline Bingham, *First Dinosaur Encyclopedia*, 2007; 'Giant Roach Fossil Found in Ohio Coal Mine', National Geographic News, http://news.nationalgeographic.com, 2001

144. Fiction.

145. Scott Hocknull and Dr Alex Cook, *Amazing Facts about Australian Dinosaurs* (book), 2006; Michael Benton, *The Penguin Historical Atlas of the Dinosaurs* (book), 1996; Dr David Norman, *The Illustrated Encyclopedia of Dinosaurs* (book), 1985; John Long, *Dinosaurs of Australia* (book) 1989

146. Michael Benton, *The Penguin Historical Atlas of the Dinosaurs* (book), 1996; Michael Benton, *Kingfisher Factfinder: Dinosaurs* (book), 2001; ABC News in Science, www.abc.net.au, 2003

147. Phil Bell and Eric Snively, 'Polar dinosaurs on parade: a review of dinosaur migration', *Alcheringa: An Australasian Journal of Palaeontology*, vol. 32 (3), 2008

148. Dinosaurs for Kids, www.kidsdinos.com, 2010; Melbourne Museum, http://museumvictoria.com.au, 2010

149. Fiction.

Sources

150. Fiction.

151. Caroline Bingham, *First Dinosaur Encyclopedia*, 2007; HowStuffWorks.com, http://animals.howstuffworks.com, 2010

152. Michael Benton, *The Penguin Historical Atlas of the Dinosaurs* (book), 1996; John Long, *Insiders: Dinosaurs* (book), 2007; Vertebrate Paleontology at Insubria University, http://dipbsf.uninsubria.it; Enchanted Learning, www.enchantedlearning.com, 2010

153. Dr David Norman, *The Illustrated Encyclopedia of Dinosaurs* (book), 1985; Michael Benton, *Kingfisher Factfinder: Dinosaurs* (book), 2001

154. 'Paleontologists Teach Medical Students about Fossil Tumors', Science Daily, www.sciencedaily.com, 2006

155. Michael Benton, *The Penguin Historical Atlas of the Dinosaurs* (book), 1996; University of California Museum of Paleontology, www.ucmp.berkeley.edu, 2005; Dinosaurs for Kids, www.kidsdinos.com, 2010

156. Fiction.

157. Melbourne Museum, http://museumvictoria.com.au, 2010; Enchanted Learning, www.enchantedlearning.com, 2010

158. Michael Benton, *The Penguin Historical Atlas of the Dinosaurs* (book), 1996; David Fastovsky and David Weishampel, *Dinosaurs: A Concise Natural History* (book), 2009

159. Fiction.

160. Luxury Watch Report, www.luxurywatchreport.com, 2009; MSNBC Digital Network, www.msnbc.msn.com, 2010

161. Michael Benton, *The Penguin Historical Atlas of the Dinosaurs* (book), 1996; Dr George Johnson's Backgrounders, http://txtwriter.com, 2008

162. *Encarta Encyclopedia* (CD), 1999; University of California Museum of Paleontology, www.ucmp.berkeley.edu, 2006; Women in Science, www.sdsc.edu, 1997

163. Dino Data, www.dinodata.org, 2010; 'Crichtonsaurus', Bob Strauss, About.com, http://dinosaurs.about.com, 2010

164. 'Attenborosaurus', Bob Strauss, About.com, http://dinosaurs.about.com.htm, 2010; The Plesiosaur Directory, http://plesiosaurnews.wordpress.com, 2010

165. Fiction.

166. 'Cryolophosaurus', Bob Strauss, About.com, http://dinosaurs.about.com, 2010; Enchanted Learning, www.enchantedlearning.com, 2010

167. Fiction.

168. 'The 10 Strangest Dinosaur Names: Dinosaur Names Even YOU Couldn't Come Up With!', Bob Strauss, About.com, http://dinosaurs.about.com, 2010; 'Colepiocephale', Bob Strauss, About.com, http://dinosaurs.about.com, 2010; Thescelosaurus, www.thescelosaurus.com, 2010

169. Fossilsmith Studios, www.triunecommunications.com.html, 2005; The Children's Museum of Indianapolis, www.childrensmuseum.org, 2010; 'Bambiraptor', Bob Strauss, About.com, http://dinosaurs.about.com, 2010

170. 'Shuvuuia', Bob Strauss, About.com, http://dinosaurs.about.com, 2010; Luis V Rey's Art Gallery: Dinosaurs and Paleontology, www.luisrey.ndtilda.co.uk; American Museum of Natural History, www.amnh.org, 2010; Enchanted Learning, www.enchantedlearning.com, 2010

171. Fiction.

172. Walking with Dinosaurs (ABC, BBC), www.abc.net.au, 1999

173. Fiction.

174. Walking with Dinosaurs (ABC, BBC), www.abc.net.au, 1999

175. Fiction.

176. Australian Museum, http://australianmuseum.net.au, 2009

177. Fiction.

178. Fiction.

179. 'China Spends Billions to Study Dinosaur Fossils at Sites of Major Discoveries', *The Washington Post* (newspaper), Washington DC, USA, 26 January 2010

180. 'Chemical Remains of Dinobird Found', EurekAlert, www.eurekalert.org, 2010

181. 'Dinosaurs: How Long Did They Live?', Guardian News and Media, www.guardian.co.uk, 2009

182. Fiction.

183. Fiction.

184. Dinosaur National Monument – National Park Service, US Department of the Interior, www.nps.gov, 2000; 'New Dinosaur Rears Its Head in National Monument', National Park Service, US Department of the Interior, www.nps.gov, 2010

185. 'Epidendrosaurus', Bob Strauss, About.com, http://dinosaurs.about.com, 2010; 'Big Bad Bizarre Dinosaurs', National Geographic, http://ngm.nationalgeographic.com, 2007; National Geographic, http://animals.nationalgeographic.com.au, 2010

186. National Geographic, http://ngm.nationalgeographic.com, 2007; Enchanted Learning, www.enchantedlearning.com, 2010

187. Fiction.

188. My Jurassic Park, www.myjurassicpark.com, 2006

189. Michael Benton, *The Penguin Historical Atlas of the Dinosaurs* (book), 1996; Caroline Bingham, *First Dinosaur Encyclopedia*, 2007; Phil Bell and Eric Snively, 'Polar dinosaurs on parade: a review of dinosaur migration', *Alcheringa: An Australasian Journal of Palaeontology*, vol. 32 (3), 2008

190. 'Giant Dinosaur Footprints Found in Argentine "Jurassic Park" ', Yahoo News, http://news.yahoo.com, 2010

191. Scott Hocknull and Dr Alex Cook, *Amazing Facts about Australian Dinosaurs* (book), 2006

192. Fiction.

193. 'Eustreptospondylus', Bob Strauss, About.com, http://dinosaurs.about.com, 2010; Walking with Dinosaurs (ABC, BBC), www.abc.net.au, 1999

194. 'Jurassic Fast Food Was a Key to Giant Dinosaurs', Science Daily, www.sciencedaily.com, 2010

195. Fiction.

196. HowStuffWorks , http://animals.howstuffworks.com, 2010; Enchanted Learning, www.enchantedlearning.com, 2010; 'Torosaurus', Bob Strauss, About.com, http://dinosaurs.about.com, 2010

197. Palaeontological Society of the Peace, www.gprc.ab.ca, 2009

198. Fiction.

199. Fiction.

200. 'Dinosaur Mummy Found With Fossilized Skin and Soft Tissues', Science Daily, www.sciencedaily.com, 2007; 'Mummified Dinosaur Found by Tyler Lyson '06 Is Ready for Its Closeup', Swarthmore College News, www.swarthmore.edu, 2007; 'Mummified Dinosaur Slowly Being Revealed', Fox News, www.foxnews.com, 2008

201. Fiction.

Sources

202. Walking with Dinosaurs (ABC, BBC), www.abc.net.au, 1999; 'How Many Insect Species Are There?', The Science Show: ABC Radio National, www.abc.net.au, 27 April 2002; 'How Many Fish in the Sea? About 20,000 Species', CTV News, www.ctv.ca, 2010; Burt Monroe and Charles Sibley, *Distribution and Taxonomy of Birds of the World* (book), 1991

203. Enchanted Learning, www.enchantedlearning.com, 2010; 'New Dinosaur Was Nut-Cracking "Parrot"', National Geographic News, http://news.nationalgeographic.com, 2009

204. Enchanted Learning, www.enchantedlearning.com, 2010; 'Styracosaurus', Bob Strauss, About.com, http://dinosaurs.about.com, 2010